John Call Dalton

History of the College of Physicians and Surgeons

In the City of New York

John Call Dalton

History of the College of Physicians and Surgeons
In the City of New York

ISBN/EAN: 9783337414580

Printed in Europe, USA, Canada, Australia, Japan

Cover: Foto ©berggeist007 / pixelio.de

More available books at **www.hansebooks.com**

HISTORY

OF THE

COLLEGE OF PHYSICIANS AND SURGEONS

IN THE CITY OF NEW YORK;

Medical Department of Columbia College.

BY

JOHN C. DALTON, M.D.,
PRESIDENT, AND PROFESSOR EMERITUS OF PHYSIOLOGY.

PUBLISHED BY ORDER OF THE COLLEGE
NEW YORK
1888

TRUSTEES

OF THE

COLLEGE OF PHYSICIANS AND SURGEONS, 1888.

HON. FREDERICK A. CONKLING
JOHN J. CRANE, M.D.
ELLSWORTH ELIOT, M.D.
ROBERT G. REMSEN, ESQ.
REV. F. A. P. BARNARD, LL.D.
SAMUEL T. HUBBARD, M.D.
THOMAS F. COCK, M.D.
WILLIAM W. HOPPIN, JR., ESQ.
GEORGE A. PETERS, M.D.
HON. GILBERT M. SPEIR
HON. EDWARD MITCHELL
WILLARD PARKER, M.D.
WILLIAM H. DRAPER, M.D.
JOHN C. DALTON, M.D., President
JAMES T. SWIFT, ESQ.
THOMAS M. MARKOE, M.D., Vice-President
CORNELIUS VANDERBILT, ESQ.
GEORGE G. WHEELOCK, M.D., Registrar
JAMES A. ROOSEVELT, ESQ.
GEORGE BLISS, ESQ., Treasurer
JOSEPH H. CHOATE, ESQ.
MORRIS K. JESUP, ESQ.
RT. REV. HENRY C. POTTER, S.T.D., Chaplain
A. BRAYTON BALL, M.D.

TABLE OF CONTENTS.

CHAPTER I.
FOUNDATION AND EARLY YEARS OF THE COLLEGE.
1807—1811.

Introduction—Origin of the College—Charter of 1807—Amendment of 1808—The faculty in 1807—Opening of the first session—The college building in Robinson street—New York City in 1807—The faculty in 1808—Methods of instruction—Removal to Magazine street—Dr. Romayne—Dr. Mitchill—Dr. Miller—Dr. Bruce—Dr. DeWitt—Dr. Macneven—First graduating exercises of the College . . . . Pages 7—28.

CHAPTER II.
THE COLLEGE IN BARCLAY STREET.
1811—1826.

Charter of 1811—Charter of 1812—Accession of professors from Columbia College—The faculty in 1814—Removal to Barclay street—The college building in Barclay street—Its enlargement—The Elgin Botanic Garden—Dr. Bard—Dr. Hosack—Dr. Mott—Dr. Francis—Dr. Post—Disputes with the County Medical Society—Change in the Board of Trustees—Disputes between the professors and the trustees—Investigation by the Regents—Further change in the Board of Trustees—Resignation of the faculty—Appointment of new professors by the Regents . . . . Pages 29—54.

CHAPTER III.
REORGANIZATION OF THE COLLEGE.
1826—1837.

The faculty in 1826—Dr. John Watts—Dr. John Augustine Smith—Dr. Dana—Dr. Stevens—Dr. Delafield—Dr. Joseph M. Smith—Dr. Beck—Financial embarrassments of the College—Perseverance against difficulties—Members of the Board of Trustees—Dr. John Kearny Rodgers—Dr. Jaques—Dr. Hamersley—Dr. Johnston—Dr. Borrowe—Dr. Cheesman—Mr. Hamilton—Mr. King—Mr. Bruen—Mr. Campbell—Mr. Allen—Mr. Troup—Mr. Boyd—Death of Dr. Dana—Appointment of Dr. Torrey—Reappointment of Dr. Mott—Removal to Crosby street . . . . Pages 55—67

CHAPTER IV.

The College in Crosby Street.
1837—1856.

Resignation of Dr. Stevens—Appointment of Dr. Parker—Dr. Watts—Dr. Gilman—Dr. John Augustine Smith—The faculty in 1843—Improvements in the College—Spring and Fall lectures—Lengthening of the college term—Establishment of college clinics—Bellevue Hospital—Legalization of practical anatomy—The Pathological Society—The Academy of Medicine . . . . Pages 68—85.

CHAPTER V.

Changes in the Faculty.
1851—1858.

Dr. Beck—Dr. Bartlett—Chair of Physiology and Pathology—Dr. Clark—Chair of Materia Medica and Clinical Medicine—Of Pathology and Practical Medicine—Of Physiology and Microscopic Anatomy—Dr. Dalton—Dr. St. John—Dr. Stevens—His resignation as president—Appointment of Dr. Cock—Of Dr. Delafield . . . . Pages 86—97.

CHAPTER VI.

Removal to Twenty-third Street.
1856—1860.

Project of removal from Crosby street—Purchase of land in Twenty-third street—Erection of the new building—Its inauguration—The faculty in 1858—Release of the College from the authority of the Regents—Union with Columbia College . . . . Pages 98—107.

CHAPTER VII.

The College in Twenty-third Street.
1856—1887.

Establishment of the Alumni Association—The Alumni prize—The Cartwright prize—The Cartwright lectures—Laboratory of the Alumni Association—The endowment fund—The laboratory fund—Prizes for undergraduates—Harsen prizes for clinical reports—Injurious effect of special prizes for undergraduates—Their abolition—Prizes for general proficiency in examination—Growth of the college clinics—Of the

Physiological and Pathological Laboratory—Changes in methods of examination—Lengthening of the term . Pages 108—135.

CHAPTER VIII.
CHANGES IN THE FACULTY.
1865—1887.

Dr. Chandler R. Gilman—Dr. Thomas—Dr. McLane—Dr. Tuttle—Dr. Joseph M. Smith as professor of materia medica—Succeeded by Dr. McLane—Dr. Edward Curtis—Dr. Peabody—Dr. Watts professor of anatomy—Dr. Sands—Dr. Sabine—Dr. Parker professor of surgery—His resignation—Dr. Markoe—Dr. Sands—Dr. Hall—Dr. Bull—Dr. St. John professor of chemistry—Succeeded by Dr. Chandler—Dr. Francis Delafield professor of pathology and practical medicine—Dr. John G. Curtis professor of physiology—Dr. Edward Delafield, eighth president of the College—Dr. Clark, ninth president—The faculty in 1887 . . . . Pages 136—152.

CHAPTER IX.
GIFTS AND BEQUESTS TO THE COLLEGE.
1875—1886.

Dr. McClelland—His bequests to the Alumni Association and to the College—Mr. James T. Swift—His gift for a physiological cabinet—Mr. William H. Vanderbilt—His gift of land and building fund—His reasons for the benefaction—His life and character—Mr. and Mrs. William D. Sloane—The Sloane Maternity Hospital—The Vanderbilt Clinic—Increased capacity of the College from these endowments . . Pages 153—164.

CHAPTER X.
THE COLLEGE IN FIFTY-NINTH STREET.
1887—1888.

Plans for the new building—Its erection and inauguration—Description of the college building—The Sloane Maternity Hospital—Its construction and organization—The Vanderbilt Clinic—Its departments—Clinical staff—The Roosevelt Hospital—Endowment and organization—Hospital Staff—New measures adopted by the College—Entrance examinations—Lengthening of the term—Changes in the curriculum—Establishment of a three years' college course . . . Pages 165—197.

BIBLIOGRAPHY.

The materials for this work have been drawn from the records of the College of Physicians and Surgeons, and also from the following:

Minutes of the Regents of the University of the State of New York, from 1791 to 1850, inclusive.

Minutes of the Medical Society of the county of New York, for 1807, 1809, 1819, 1820, and 1825.

ROMAYNE.—Report and address delivered by the President to the Medical Society of the county of New York. Published by the Society. New York, 1807.

MITCHILL.—Picture of New York, or the Traveller's Guide through the commercial metropolis of the United States. By a gentleman residing in this city. New York, 1807.

HOSACK.—Statement of facts relative to the Elgin Botanic Garden, New York, 1811.

Exposition of the transactions relative to the College of Physicians and Surgeons in the city of New York. New York, 1812.

HOSACK.—Observations on the establishment of the College of Physicians and Surgeons in the city of New York and the late proceedings of the Regents of the University relative to that institution. New York, 1811.

HOSACK.—Historical sketch of the origin, progress and present state of the College of Physicians and Surgeons of the University of the State of New York. (With a view of the College), New York, 1813;

also in the American Medical and Philosophical Register. New York, 1814.

Account of the life and character of Edward Miller, M.D., American Medical and Philosophical Register. 1812.

Memoir of Archibald Bruce, M.D., in the American Journal of Science. New York, 1818.

Obituary notice of Nicholas Romayne, M.D., in the Medical Repository. New York, 1818.

STEARNS.—Annual address before the Medical Society of the State of New York. Albany, 1818.
Obituary notice of Benjamin De Witt, M.D., in the Medical Repository. New York, 1819.
MITCHILL.—Discourse on the life and character of Samuel Bard, M.D.; New York, 1821.
MACVICKAR.—Life of Samuel Bard, M.D.; New York, 1822.
HOSACK.—Inaugural Discourse at the opening of Rutgers Medical College. New York, 1826.
BECK.—Obituary notice of James Freeman Dana, M.D.; in the New York Medical and Physical Journal. New York, 1827.
THACHER.—American Medical Biography; Boston, 1828.
SMITH, J. A.—Eulogium on the late Wright Post, M.D.; in the New York Medical and Physical Journal, 1828.
Obituary notice of Samuel L. Mitchill, M.D.; in the New York Medical Journal, 1831.
FRANCIS.—Reminiscences of Samuel Latham Mitchill, M.D., New York, 1859.
MOORE.—Memoir of John Watts, M.D., New York, 1831.
WILLIAMS.—American Medical Biography. Greenfield, 1845.
MOTT.—Reminiscences of medical teachers and teaching in New York. New York, 1850.
GILMAN.—Sketch of the life and character of John Brodhead Beck, M.D.; in the New York Journal of Medicine, 1851.
FRANCIS.—Old New York, or Reminiscences of the past sixty years. New York, 1858.
MOTT.—Eulogy on the late John W. Francis, M.D. New York, 1861.
GROSS.—Lives of eminent American Physicians and Surgeons. Philadelphia, 1861.
BLATCHFORD.—Our Alma Mater fifty years ago. Oration delivered before the Alumni Association, March 14th, 1861.
ADAMS.—Memoir of Jacob Harsen, M.D., read before the New York Academy of Medicine, June 1st, 1864.
POST.—Eulogy on the late Valentine Mott, M.D.; delivered before the N. Y. Academy of Medicine, November 27th, 1865.
ADAMS.—Discourse commemorative of the life and character of Alexander H. Stevens, M.D.; New York, 1871.
Obituary notice of John McClelland, M.D.; in the New York Medical Register for 1875-6.
LAMB.—History of the city of New York. New York, 1880.

CHAPTER I.

INTRODUCTION.

FOUNDATION AND EARLY YEARS OF THE COLLEGE.

1807–1811.

The College of Physicians and Surgeons has reached a point in its history when it seems desirable to collect, in a permanent and accessible form, the records of its growth. Now in the eighty-second year of its existence, it occupies a position of strength and security, guaranteed by continuous progress in the past and the promise of greater prosperity in the future. It is the oldest existing medical school in the State of New York, and one of the four in the United States which have lasted for more than three quarters of a century. It bears upon its rolls the names of over four thousand five hundred graduates; and its alumni have always shown a deep solicitude for its welfare. There are none now living whose personal recollections connect them in any degree with the early years of the institution; and for those of the present generation its annals have all the attraction of historic interest. In some respects, moreover, they may yield instruction as well as entertainment. The present condition of the College is by no means wholly due to events or influences of our own day. Its destiny has been shaped by many different causes following each other in long succession, the more recent always connected in some measure with those which have gone before. Every improvement in its organization or operation has been accomplished by repeated and persevering endeavor, by gradual enlargement rather than by sudden or forcible expansion. It has had its trials and vicissitudes, its failures as well as its successes; and we may surely learn something from experience, to avoid the repetition of past mistakes. A difficulty once removed, by means which have proved effectual, need not again be encountered if the institution adhere to its policy of a discriminating and judicious conservatism. On the other hand, it has ever been ready to alter its customs and regulations when they have become useless or inappropriate and no longer fulfil the purpose for which they were designed. Its methods have varied to meet the necessity of the times; its aims and objects have always been the same.

The College took its origin, in the first decade of the century, from a spontaneous movement of the profession in the city of New York for the cultivation and improvement of medical science and art. In the year 1807 the Medical Society of the County of New York adopted a memorial to the legislature, setting forth the desire of its members to "promote the progress of medical knowledge," and to give "encouragement and protection" to the pursuit of medical science; and expressing the belief that "their usefulness would be extended in promoting the public good and the improvement of their profession" by their incorporation as a college under the auspices of the State University. They accordingly petitioned to be so incorporated "under the direction, inspection, and patronage of the Regents of the University," in such manner as might be deemed proper "for the public good, and the promotion and improvement of the medical profession and the sciences connected therewith."

At the same time a memorial was addressed to the Regents of the University informing them of the efforts of the Society for the diffusion of science and the improvement of the profession, and declaring that these efforts "would be more successful if they were directed under the patronage of the Regents," and praying them to "favor the views of the said Society."

This action met with a prompt response from the Regents of the University; and within a month thereafter the members of the Society were duly incorporated as a college, by a charter bearing date March 12th, 1807. The charter was granted by the Regents under authority conferred on them by a previous act of the legislature.

The Medical Society of the County of New York was then in the second year of its existence, and numbered one hundred and thirty-nine members; embracing all the legally qualified practitioners in the vicinity.

Its constituent meeting, July 1st, 1806, is called, in the printed report of the proceedings, a "meeting of the physicians and surgeons of the city and county of New York." In the address of its presiding officer on the following day it is spoken of by the same designation; and when, beside their organization as a society, its members were also incorporated as a college, the institution so established was entitled the COLLEGE OF PHYSICIANS AND SURGEONS in the city of New York. The members of the society became, by the charter, members of the college. They were empowered to elect annually its president, vice president, registrar, treasurer and censors; to make by-laws, rules and regulations relative to its affairs and property; and to direct the disposal of moneys in the hands of the treasurer. They were responsible for their administration only to the Regents of the University, to whom was reserved the appointment of professors and the right of conferring degrees.

The College was therefore, in its primary organization, a creation of the Medical Society of the County of New York; and the body of its trustees or members comprised all the members of that Society. This forms one of its most honorable claims to distinction. It represented the best endeavors of the profession for the diffusion of medical knowledge and a better medical education. It embodied their hopes for the immediate needs of the time, and laid the foundation for further improvement in the future.

But an educational institution can hardly be administered successfully by so large a body of managers as the whole membership of the county medical society. However sincere their intention, they could not all possess the requisite knowledge nor command the necessary time for the regulation of its affairs; and in point of fact at their first meeting, to complete the college organization, May 5th, 1807, only sixty-three members were present. Other defects in the charter became apparent on trial. Under its provisions the executive officers of the College were to be elected annually. But these officers require, for the due performance of their functions, the experience and familiarity of continuous service; and the business of the institution would be liable to derangement if they were too frequently changed.

These imperfections seemed so obvious that in the following year the College presented to the Regents a request for certain changes, which were considered "important to the stability and usefulness of the institution;" and the charter was consequently amended by an ordinance passed March 3d, 1808.

The alterations introduced by this amendment were two-fold. First, the officers of the College, instead of being elected annually by the corporation, were appointed by the Regents, thus giving greater stability to the organization; and secondly, all members of the medical society who wished to serve as trustees or members of the College were required, as a condition, to declare in writing their acceptance of the trust, and that they would, "to the best of their abilities, endeavour to promote the usefulness of the said College, and faithfully execute the several duties required of them." By this means the institution was relieved of its doubtful or indifferent members, and was entrusted to those who had faith in its destiny and would give it the guaranty of their favor and support.

In the mean time the College had elected its officers, adopted a code of by-laws, and had been provided with a corps of professors and lecturers, ready to carry on the business of instruction. They were as follows:

THE FACULTY IN 1807.

NICHOLAS ROMAYNE, M.D., *President, and Lecturer on Anatomy.*

SAMUEL L. MITCHILL, M.D., *Vice President, and Professor of Chemistry.*

EDWARD MILLER, M.D., *Professor of the Practice of Physic and Clinical Medicine.*

DAVID HOSACK, M.D., *Professor of Materia Medica and Botany, and Lecturer on Surgery and Midwifery.*

ARCHIBALD BRUCE, M.D., *Registrar, and Professor of Mineralogy.*

BENJAMIN DE WITT, M.D., *Professor of the Institutes of Medicine and Lecturer on Chemistry.*

JOHN AUGUSTINE SMITH, *Adjunct Lecturer on Anatomy.*

All of these gentlemen were residents of the city and members of the County Medical Society, with the exception of John Augustine Smith, who came from Virginia to accept his appointment in the College. He was a member of the Royal College of Surgeons of London, where he had completed his medical education. He afterward received the honorary degree of M.D. from the College of Physicians and Surgeons where he was serving as professor.

At a meeting of the College in June, 1807, the president, professors, and lecturers were constituted a "Senatus academicus," or a sort of standing committee, to "correspond with the medical societies in the several counties in the State," to make regulations for the teaching department, to ascertain what branches of medical science "usually taught in the most respectable universities" were still unprovided for, and to nominate fit and proper persons to fill such vacancies.

Under this authority a circular was issued to the presidents of the different county medical societies, informing them of the establishment of the College and of the intended course of instruction. It is evident from various sources that the College was intended to be an institution of wide and general usefulness to the profession, embracing, as one of its functions, that of systematic medical teaching. The circular declares that "it is the principal object of this new Institution to assist the progress of medical science in every part of the State of New York;" and that its members consider " the cultivation of correspondence and intimate connection with the Medical Society of the State, and the Medical Societies of the several Counties as one of their most important duties." It also announces that "under the direction and patronage of the Regents the College of Physicians and Surgeons have instituted a School of Physic, which it will be their unremitting endeavour to render equal, in extent, comprehensiveness, and accuracy of instruction, to the most distinguished universities of Europe;" and that they have procured, "in a central part of the city," a commodious building, where apartments will be fitted up for the use of the lecturers and students. The arrangements were completed in accordance with this plan, and the first course of lectures was duly opened on Tuesday, November 10th, 1807.

But such an institution is not set on foot without the expenditure of money. Many outlays must be incurred for indispensable objects before there can be any source of revenue. The annual rent of the house secured for the use of the College was $800. The necessary repairs, fittings, and furniture, to make it available for its new purpose, cost $730. Moreover, there were charges for fuel, printing, wages, anatomical material, chemical apparatus and supplies, and other incidental requisites, amounting during the year to $1,120 additional. How were these unavoidable expenditures to be met? In the first place, there were "contributions" from various members of the College, in sums of from ten to twenty dollars each, amounting in all to $230. Secondly, Dr. Romayne gave his note of hand for $5,000, which was used by the College as security to obtain advances of money from the Manhattan Bank. Subsequently, Dr. Romayne was joined in this guaranty by Drs. Miller and Bruce, and further sums were added to it until it amounted, in 1810, to over $8,000. Meanwhile a grant was made to the College by the legislature of an interest in the "literature fund lotteries," originally established for the benefit of Union College in Schenectady. This grant was to yield twenty thousand dollars, in successive instalments of five thousand dollars each. The first of these instalments was paid in 1810, and was used to liquidate a part of the above debt; but the joint obligation of Drs. Romayne, Miller, and Bruce was not fully discharged until December, 1813.

The "central part of the city," in which the College was inaugurated and where its first sessions were held, was Robinson street. It was a short street, running west from Broadway to the grounds then occupied by Columbia College; and formed a portion of what is now Park Place.* Even its present designation has long been a misnomer, since the disappearance of the Park from its eastern extremity and its extension as a street in the opposite direction to the North River. The college building was at number 18, on the south side of the street. Probably not a single feature of the locality, as it then was, exists

* The name of Robinson street was changed to Park Place in 1813.

to-day; but for the friends of the institution it must always retain a certain mysterious interest, as the earliest domicile of the College of Physicians and Surgeons.

At that time the population of New York was but little more than eighty thousand. Most of the city was below Chambers street. The wealthier residences were at the lower end of Broadway, about the Battery and Bowling Green, with the shops in the upper part of the same street. Broadway was paved only to the neighborhood of Canal street, beyond which it continued as a road. Canal street itself existed only on paper, and was represented by a swamp and a sluggish stream, crossed by a bridge at the intersection of Broadway. The New York Hospital was in an open space on the west side of Broadway, between the present Duane and Worth streets. Its approach from Broadway was bordered with elms; and it had on one side a kitchen garden, to supply it with vegetables. It was three stories in height; and the cupola on its roof commanded an extensive view, embracing "the entire city," as well as "the harbour and country beyond, to a great distance." The Park, in the upper portion of the city, occupied the triangular space between Broadway and Park Row, now covered in great measure by the Post Office building and the adjoining thoroughfare. It was "planted with elms, planes, willows, and catalpas." The space was enclosed with a wooden paling, and the surrounding foot-walk "encompassed by rows of poplars." The water-supply of the city was from wells and pumps, usually situated in the middle of the street; the water being distributed thence in casks by wagons to the house doors. There was a stage route from Wall street to Greenwich village, about two miles distant, the present vicinity of Christopher street. The ferries to Brooklyn and the Jersey shore were served by row boats and small sailing craft. There were neither Croton water-works nor gas companies; and none of the streets were occupied by telegraph lines or elevated railroads.

But notwithstanding this contrast with the appearance of the city at present, it would be a mistake to suppose that there was anything that could be called primitive in its people or their mode of life. They occupied a smaller area and lived at the beginning of the century; but except for the differences of space and time, they were as busy, enterprising, luxurious and progressive then as now. The population of the city had increased over thirty per cent. since 1800; and in certain localities the land was said to have tripled in value within twenty years. Mr. John Lambert, an English writer who visited this country in 1807, found every indication of prosperity. In the business part of the town "all was life, bustle and activity." The houses in Broadway were "lofty and well built;" the shops "large and commodious, well stocked with European and India goods, and exhibiting as splendid and varied a show in their windows as could be met with in London." The manner of living, among the wealthy and professional classes, seemed "little inferiour to that of Europeans;" their houses being "furnished with everything agreeable or ornamental," and "fitted up in the tasteful magnificence of modern style."* There were seven or eight daily newspapers, and a medical quarterly edited by two of the most prominent members of the profession.

Such was the condition of the city and the times when the College of Physicians and Surgeons came into existence. Its friends believed that a promising field of prosperity and usefulness lay before it, and the immediate result was not disappointing. Its first course of lectures was attended by fifty-three students. Early in the following year it appeared to the Regents of the University that the College had commenced its business in a manner "to answer all the expectations entertained in its establishment;" and they recommended it, in their report to the legislature, as an institution "important to the welfare of the people of the State." At the second session its class numbered seventy-six, and at the third eighty-two.

After the first year the faculty was completed by the permanent appointment of professors in place of lecturers, giving it the following organization:

---

* This *modern style* was the revival of classical taste in architecture and house decoration, which originated in France under the consulate and the empire, and extended thence to England and the United States. It continued in vogue, in this country, for about twenty years.

THE FACULTY IN 1808.

Nicholas Romayne, M.D., *President, and Professor of the Institutes of Medicine.*

Samuel L. Mitchill, M.D., *Vice President, and Professor of Natural History and Botany.*

Edward Miller, M.D., *Professor of the Practice of Physic and Clinical Medicine.*

Archibald Bruce, M.D., *Professor of Mineralogy and Materia Medica.*

Benjamin De Witt, M.D., *Professor of Chemistry.*

John Augustine Smith, *Professor of Anatomy and Surgery.*

William James Macneven, M.D., *Professor of Obstetrics and the Diseases of Women and Children.*

Such a course of instruction was intended to answer the requirements of a thoroughly liberal medical education; including, in addition to the ordinary topics, the auxiliary branches of natural history, botany and mineralogy. The "institutes of medicine" comprised physiology and hygiene, the general doctrine of the causes and symptoms of disease, and general therapeutics. Mineralogy was taught by the professor of materia medica, and anatomy and surgery were also confided to a single chair. The distribution of subjects was therefore unlike that generally adopted at the present day. The regular lecture term was of four months' duration.

In some other respects the methods of that time were different from those to which we are now accustomed. The main business of the session was approached with a certain formality, befitting the professional dignity of the occasion. The first day was devoted to an introductory address by the president, at twelve o'clock. Afterward the professors followed in turn, on successive days, at the same hour, each with an introductory in his own department; so that on the whole an entire week was taken up with preliminary medical literature. But the regular course once fairly opened, the work of instruction was carried on with industry and zeal. It appears from the programme for 1808 that five lectures were given in the College every day. Some of the professors lectured four times a week, others daily throughout the session. Both their time and that of their pupils must have been fully occupied.

It is also noticeable that measures were taken from the very beginning, to secure the advantages of hospital instruction. At a meeting of the College in June, 1807, it was *Resolved*, "that the Senatus academicus be empowered, on the part of the College, to confer with the Governors of the New York Hospital relative to the promotion of medical education;" and later in the same year it was announced that the students would "have an opportunity of attending the practice of Dr. Miller at the New York Hospital, the Governors of that Institution having with great liberality made arrangements for that purpose." The hour for clinical instruction at the hospital was from twelve to one o'clock. Visits were also made with Dr. Macneven at the almshouse, then situated in Chambers street on the site of the present city court house.

The house in Robinson street was one hired for temporary use, and at the end of the second year the College was removed to a building of its own in Magazine street. This was a street extending east from a point in Broadway opposite the grounds of the hospital. It afterward became the upper end of Pearl street, with which it was united in 1811. The property consisted of a lot twenty-five feet in width by one hundred feet in depth, having upon it a dwelling house, probably of two or two and a half stories. It was purchased in 1807 by Dr. Romayne, who at first held it in trust, the title being afterward formally transferred to the College. It was on the south side of the street, near Broadway, and corresponded with the present number 553 Pearl street. After being fitted up for the reception of the College, it was occupied in November, 1809.

This was the history of the institution for the first few years of its existence. During that time it numbered in its faculty several members of marked character and ability.

Foremost among these was NICHOLAS ROMAYNE, the most active man in the organization of the College, and its first president. He was the delegate who obtained its charter from the Regents of the University; he pledged his personal credit to provide it with funds; and he delivered for three years the lectures on the institutes of medicine. He was a little over fifty years of age, and already a successful teacher of private pupils in nearly all the departments of medicine. He was elected president of the Medical Society of the County of New York at its organization in 1806, and was president of the State Medical Society in 1809, 1810 and 1811; and he had been zealous in procuring from the legislature a variety of laws for the benefit of the profession. He was a man of large stature, but easy and graceful motion; of vigorous and cultivated mind, active ambition and persistent energy; and of a disposition always ready to accept the responsibilities of the occasion. If the College can be said to have been established through the special exertion and influence of any one man, Dr. Romayne must undoubtedly be regarded as its founder.

Equally notable was Dr. *Samuel L. Mitchill*, vice president of the College, senator of the United States, professor of Chemistry and Natural History. He was between forty and fifty years of age, and of wide reputation as a man of talent and accomplishments. According to his biographers he was a kind of "human dictionary," whose opinion was sought by schemers and inventors of every grade, and who could be consulted with profit on any question of science, history or politics. He was equally distinguished for his learning and originality, and for his "hospitality to new ideas." He could discourse in turn on a Babylonian brick, meteoric stones, the theory of chemical combination, the construction of a wind-mill, the fishes of North America, or the geology of Niagara

Falls. He was one of the founders and editors of the New York "Medical Repository," the earliest medical periodical in the United States, and he was for twenty years indefatigable in contributing to its success. He

SAMUEL L. MITCHILL, M.D.

Vice President of the College, 1807-1811. Senator of the United States, 1804-1809. Professor of Chemistry, 1807-1808; of Natural History and Botany, 1808-1820; and of Botany and Materia Medica, 1820-1826. From an engraved copy of the portrait by Jarvis, now in the library hall of Columbia College; painted about 1815.

obtained from Congress the appropriation for the defences of New York harbor; he aided De Witt Clinton in his project for the Erie canal, and was the orator of the day at the ceremony of its inauguration; he believed in Robert Fulton's idea of steam navigation, and went on the trial trip of his first steamboat to Albany. Disinterested, patriotic, engaging and communicative, he was an influential character in the creation and development of American science.

Dr. *Edward Miller*, professor of Practice and Clinical Medicine, represented in his day the best type of the learned and skilful practitioner. Of liberal education and classical tastes, he kept pace with the advance of professional knowledge; and his agreeable manners were combined with an integrity of purpose universally acknowledged. His friend, Dr. Mitchill, says of him that "his head was a treasury of information, and his heart a mine of benevolence." He was associated with Dr. Mitchill in the editorship of the "Medical Repository," and was visiting physician and clinical lecturer at the New York Hospital. With a large and lucrative practice, he possessed to an unusual degree the attachment of his patients, the esteem of his colleagues, and the confidence of the public. It is said that the concourse at his funeral, in 1812, was larger than had ever been seen in New York on a similar occasion, except in the single case of Alexander Hamilton.

Dr. *Archibald Bruce*, registrar of the College and professor of Mineralogy and Materia Medica, was younger than most of his colleagues, but well known for his scientific attainments. After graduating at Columbia College, he pursued the study of medicine at Edinburgh, where he received the degree of M.D. in 1800. He then visited various parts of Europe, spending his time in the study of mineralogy, and in collecting a valuable cabinet of specimens which he brought with him on his return to New York. He soon afterward commenced the publication of the *American Journal of Mineralogy*, the first purely scientific journal in this country, and the immediate predecessor of Silliman's "American Journal of Science." In conjunction with his former preceptor, Dr. Romayne, he was active in establishing the incorporated medical societies which did so much for the security and improvement of the profession in the County and State of New York. In a biographical notice, published in Silliman's journal in 1818, it is said that Dr. Bruce's ruling passion was love of science; his special attention being devoted to mineralogy. On this subject he became a "focus of information;" and he was particularly assiduous in bringing to light the mineral resources of the United States. He was social in his habits and disposition, ever forward to promote scientific interests, and an earnest and successful worker in his chosen field of knowledge.

Dr. *Benjamin De Witt*, professor of the Institutes of Medicine, also lectured on Chemistry in 1807 during Dr. Mitchill's absence at Washington as member of the Senate. The following year he was appointed professor of that branch, Dr. Mitchill assuming the more congenial department of Natural History. Dr. De Witt was then established in the city as a practitioner, having graduated in medicine at the university of Pennsylvania in 1797. He was the previous occupant of the house in Robinson street selected for the College, where his name appears as a resident in the city Directory for 1807. He afterward became vice president of the College, and one of the censors of the State Medical Society. He was co-editor and contributor to the New York "Medical and Philosophical Journal and Review," a semi-annual publication issued in 1809, 1810 and 1811. In 1815 he was appointed Health Officer of the port of New York, and continued to perform the duties of that office till his death in 1819.

Dr. *William J. Macneven*, appointed professor of Obstetrics in 1808, was of Irish birth but received his education in Germany and graduated in medicine at the university of Vienna in 1783. After commencing practice in Dublin, he joined the society of "United Irishmen" and took part in the rebellion of 1798, for which he was arrested and imprisoned until 1802. After his liberation he spent a year or two in France; coming to this country in 1804, at the same time with Thomas Addis Emmet, his friend and political associate. He at once entered on the practice of medicine in New York, where his personal character, as well as his professional and literary qualifications, secured for him a cordial recognition. He was the author of a treatise on the *Use and Construction of the Mine*

Auger, a translation from the German, London, 1788; *Exposition of the Atomic Theory*, New York, 1819; and an edition of *Brande's Manual of Chemistry*, with notes and emendations, New York, 1821. He was also associated with Dr. De Witt, and afterward with Dr. John Augustine Smith, in the editorship of the "Medical and Philosophical Journal and Review."

At the end of the fourth session, were held the first graduating exercises of the College. After the candidates had been privately examined by the professors, a list of those found qualified was sent to the Regents of the University, in whom resided the authority for issuing the diplomas. Subsequently each candidate submitted to the faculty his graduating thesis, written "either in the English, French, or Latin language," which he was required to defend at a special public examination; and finally, all these preliminaries being fulfilled, the degrees were conferred and the diplomas delivered at a public Commencement on Wednesday, May 15th, 1811.

The exercises on this occasion were somewhat elaborate, the programme having been arranged beforehand by a committee appointed for that purpose. At half-past ten in the forenoon a procession was formed at the City Hall, consisting of "students of medicine, candidates for graduation, Members of the College, the Professors, President and Vice President, the Trustees of Columbia College, the Chancellor and Regents of the University, the Reverend

Clergy of different denominations, Physicians, Gentlemen of the bar, and Strangers of distinction." The procession, headed by the janitor of the College, moved from the City Hall to the Brick Presbyterian Church,* which it entered "in inverted order." The president ascended the pulpit, the Regents of the University seating themselves on his right and the professors on his left, upon a stage erected for the purpose; the candidates for graduation occupying seats in the body of the church. After a prayer, of prescribed form and considerable length, the candidates rose from their seats, passed into the aisle and remained standing, while the president asked the assent of the faculty and of the Regents to their graduation, and administered to them in a body the Hippocratic oath. The first candidate was then called upon the stage, to inscribe his name in the college Album; after which, having "his hands embraced by those of the president," that officer pronounced the Latin formula creating him Doctor in Medicine, and delivered to him his diploma. This ceremony having been repeated with all the graduates in succession, they listened to a charge from the vice president, and the exercises were concluded with prayer.

The graduates at this Commencement were eight in number, of whom five had attended the first session of the College in 1807-8.

---

* This church was in the triangular space between the east side of the City Hall Park, Nassau and Beekman streets. It was built in 1767. and demolished in 1857.

CHAPTER II.

THE COLLEGE IN BARCLAY STREET.

PERIOD OF DISSENSION.

1811–1826.

The four years from 1807 to 1811, during which time the organization and policy of the College remained substantially the same, may be said to embrace the first chapter of its history. The second opens in 1811, when Dr. Samuel Bard was appointed president in place of Dr. Romayne, and a change was effected in the constitution of the College by a supplementary charter. Soon afterward all the existing modifications and amendments were consolidated in a new charter, granted by the Regents of the University in 1812. By this instrument the government of the institution was vested in a Board of Trustees, limited in number to twenty-five, of which the professors, president, vice president and treasurer were appointed members. The remaining vacancies were to be filled by the Regents at their discretion; but only a few additional appointments were made at that time, leaving the Board mainly in the hands of the professors and executive officers.

A plan was also set on foot for bringing into the faculty the medical professors of Columbia College. In this institution medical lectures had been given, by teachers of acknowledged ability, for nearly twenty years. But the classes in attendance had never been large; and the whole number of graduates, from 1793 to 1813, was only thirty-five. It was already evident that the College of Physicians and Surgeons had in it the elements of success; and it was thought that by combining in a single faculty the teachers of both schools, the most complete and efficient course of instruction might be provided. This was soon after accomplished by the mutual consent of the two Boards of Trustees; the medical lectures in Columbia College being discontinued, and its professors passing into the faculty of the College of Physicians and Surgeons. Early in 1814 the change was formally ratified by the Regents of the University, and the organization of the faculty became as follows:

THE FACULTY IN 1814.

SAMUEL BARD, M.D., *President.*

BENJAMIN DE WITT, M.D., *Vice President, and Professor of Natural Philosophy.*

WILLIAM J. MACNEVEN, M.D., *Professor of Chemistry.*

SAMUEL L. MITCHILL, M.D., *Professor of Natural History and Botany.*

JOHN AUGUSTINE SMITH, M.D., } *Joint Professors of Anatomy, Physiology and Surgery.*
WRIGHT POST, M.D.,

DAVID HOSACK, M.D., *Professor of the Theory and Practice of Physic.*

WILLIAM HAMERSLEY, M.D., *Professor of Clinical Medicine.*

JOHN C. OSBORN, M.D., *Professor of Obstetrics and the Diseases of Women and Children.*

JAMES S. STRINGHAM, M.D., *Professor of Legal Medicine.*

VALENTINE MOTT, M.D., *Professor of the Principles and Practice of Surgery.*

JOHN W. FRANCIS, M.D., *Registrar, and Professor of Materia Medica.*

The additional members thus introduced into the faculty were the previous medical professors of Columbia College, excepting Dr. John W. Francis, who was an alumnus of the College of Physicians and Surgeons, having graduated at its first Commencement in 1811. The new professors were at once appointed to the Board of Trustees, of which their colleagues were already members under the charter of 1812.

This completed the change in the constitution of the College, and made an essential difference in its mode of operation. In effect, the professors became the governing, as well as the teaching body of the institution. As they were often sufficient in number to form a majority of the Board, or even a quorum by themselves, and as they were necessarily more conversant than most of their associates with the situation of affairs, their influence must predominate in shaping the policy and regulating the concerns of the College.

One of the first enterprises, under the new regime, was that of obtaining a different site and building for the College; a committee having been appointed for that purpose in January, 1813. The reason assigned for this action, in the annual report to the Regents, was that the building in Pearl (formerly Magazine) street was not only small and insecure, but "ineligi-

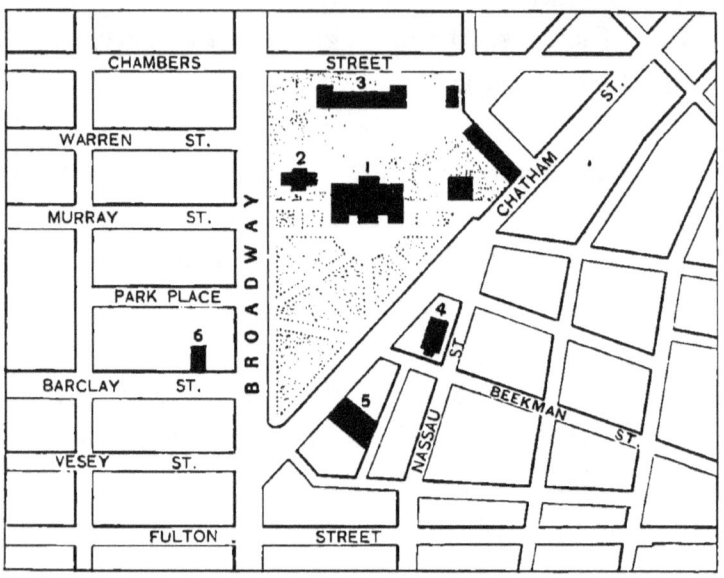

THE CITY HALL PARK AND VICINITY,

in the early part of the century; from an engraved Map of the City of New York, by Thos. H. Poppleton, city surveyor, 1817. In the Whitney collection of engravings, Columbia College Library.

1, City Hall; 2, Bridewell; 3, Old Almshouse; 4, Brick Presbyterian Church; 5, Park Theatre; 6, College of Physicians and Surgeons.

bly situated." As this ineligible situation was in close proximity to Broadway and within a few hundred yards of the New York Hospital, it seems likely that a few years' longer experience would have demonstrated its advantage as a location for the College, and justified the wisdom of its former purchase. But

THE COLLEGE BUILDING, NO. 3 BARCLAY STREET.
1813–1817.

From an engraving in the *American Medical and Philosophical Register*,
New York, 1814.

at that time it may have appeared too remote from the thickly settled portion of the city, and was perhaps inconvenient from the want of pavement and sidewalks. At all events, the recommendation of the committee was adopted, the Pearl street property was sold, and the College purchased the lot and building at number 3 Barclay street. This lot was on the north side of the street, near Broadway, and measured twenty-five feet in width by seventy-five feet in depth. The building, which was originally a brick store-house, twenty-five feet wide by thirty-eight feet deep, was so altered and repaired as to convert it into a medical college with two lecture rooms. It was three stories in height, with a terminal balustrade and a cupola, surmounted by a statue of Apollo, to indicate the scientific and medical character of the institution. It was occupied at the opening of the seventh session, on the first Monday of November, 1813.

These provisions for greater space and more varied instruction were soon followed by a larger attendance. At the eighth session (1814–15) the class numbered one hundred and twenty-one; at the ninth (1815–16), one hundred and forty-eight; and at the tenth (1816–17), one hundred and ninety-two. This increase, which was quite beyond all former anticipations, made it necessary to provide still further accommodation; and in 1817 the building was doubled in size by extending it over the adjoining lot on the west, giving it a frontage of fifty feet on Barclay street. It was fitted up with three lecture rooms; a chemical lecture room on the first floor; a hall, or general lecture and audience room, on the second floor; and an anatomical theatre on the third floor. Thus enlarged, the building continued to serve as the domicile of the College for the next twenty years.

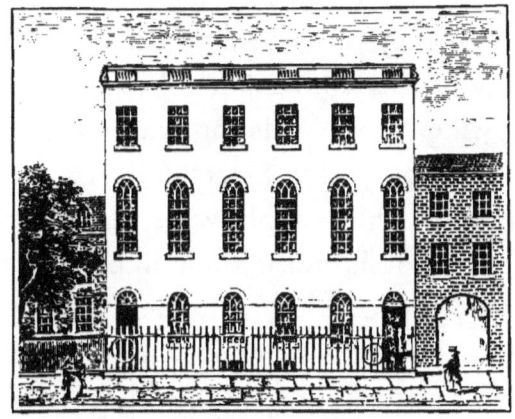

THE COLLEGE BUILDING IN BARCLAY STREET,
after its enlargement.
1817-1837.
From a print in the *Picture of New York and Stranger's Guide*,
New York, 1828.

Soon after the completion of these alterations an additional lot of land, immediately north of the College, extending through to Park Place, was secured by the purchase of its unexpired lease of twenty-four years; thus affording access to the College from the rear. At a meeting of the Trustees in October, 1817, a resolution was adopted declaring it to be " of the greatest importance to the safety and convenience of this school to possess the said lot;" and, the necessary funds having been advanced by the professor of anatomy, Dr. Wright Post, the purchase was made and reported at an adjourned meeting of the Board in the following month. At the same time it was declared expedient to erect upon this lot, in the rear of the College, "a building to answer the purposes of a stable;" and a committee was appointed to carry the plan into execution.

The expediency of securing an additional entrance for the "safety and convenience" of the College was no doubt owing to the existing exigencies and difficulties in the supply of subjects for dissection. The legal sources of this supply were at that time very inadequate; and it is more than likely, from various allusions, that the building intended to "answer the purposes of a stable" was mainly useful in connection with nocturnal expeditions in the interest of the anatomical department.

Another acquisition, then thought to be very important, was that of the "Elgin Botanic Garden." This garden was the property of Dr. Hosack,* who had purchased the land from the city, and placed it under cultivation as a botanical preserve. In the *American Medical and Philosophical Register* for July, 1811, it is described as an enclosure of about twenty acres, "on the Middle Road," † a little over

---

* It was named from the town of Elgin, in the north of Scotland, the birthplace of Dr. Hosack's father.

† This was a road running in a northerly direction, above the present site of Madison Square, in the region between the "Boston Post Road" on the east and the "Bloomingdale Road" on the west.

three miles from the city; and is illustrated by an engraved picture, showing its lawns, foot-paths, shrubs, flowers, trees, and conservatories. Its location is now included between Fifth and Sixth Avenues, from Forty-seventh to Fifty-first streets; the valuable leasehold property of Columbia College.

Dr. Hosack, finding the garden too expensive an establishment for private maintenance, proposed to transfer its ownership to the State of New York, for public instruction in botany and materia medica; and the County Medical Society adopted a memorial setting forth its advantages as an aid to medical education, and recommending its acquisition for that object. The proposal was agreed to. The Botanic Garden was purchased, under an act of the legislature, for seventy-four thousand dollars, and assigned to the keeping of the Regents of the University, as the guardians of educational interests in the State. The Regents, having no funds especially provided for maintaining the garden, placed it under the care of the College of Physicians and Surgeons, "to be by them kept in a state of preservation, and in a condition fit for all the medical purposes, free of expense, under the immediate inspection of the regents resident in the city of New York, and that the said garden be at all times open to the admission of such medical students as may resort thereto for the purpose of acquiring Botanical Science." In the college circular for 1811 it was announced that "the Botanic Garden having been purchased by the State and placed under the direction of the College, the students of Botany will have an opportunity of visiting it whenever they think proper, and of examining the many rare and valuable plants which it contains."

But no sooner was this desirable establishment in the possession of the College, than it became a source of continual annoyance. The cisterns, conservatories and heating flues needed immediate repair. A gardener was appointed under contract, to keep the grounds in proper condition; and there were disagreements between him and the College, as to how far each had fulfilled their respective engagements. Successive reports, for the next few years, showed the garden to be "much deteriorated," fences and cisterns out of repair, roads and paths in bad condition, and many of the plants lost or damaged.' There seemed little prospect of making the enterprise a success; and moreover it was found, no doubt, that medical students in Barclay street had something else to do than to make excursions to the Botanic Garden, for the purpose of examining "rare and valuable medicinal plants." Finally, the ownership of the land having been transferred, under certain conditions, by act of the legislature, to Columbia College, a committee was appointed, in 1816, to " wait on the Trustees of Columbia College and deliver to them the possession of the said Botanic Garden."

This failure of the garden as a means of medical instruction can hardly be attributed to any fault of the College. The most promising plans for improvement sometimes turn out differently from what was anticipated; as they are liable to be affected by various unforeseen conditions, the operation of which can only be learned from experience. It is probable that none of those formerly interested in the Botanic Garden had any idea of the service it would finally render in the cause of education, as a source of princely revenue to Columbia College.

The second president of the College of Physicians and Surgeons, DR. SAMUEL BARD, was everywhere regarded with esteem and consideration. He had formerly been professor of the practice of physic in Columbia College, dean of the medical faculty in the same institution, and visiting physician to the New York Hospital. At the time of his appointment as president he was sixty-nine years of age and no longer a resident of the city; having retired from practice, and spending most of the time at his country seat, at Hyde Park on the banks of the Hudson. It does not appear that he performed any active service in the College, beyond that of presiding at commencement exercises and delivering an occasional address. But his name was considered a valuable guaranty for the character and prospects of the institution, and he remained its president until his death in 1821. His successor in the presidential office was Dr. Wright Post, the professor of anatomy and physiology.

The most prominent man in the affairs of the College at this time was *David Hosack*. He was appointed professor at its organization in 1807, but for some reason did not retain his position after the first year. He returned to it in 1811, and became mainly instrumental in bringing into the faculty his former associates of Columbia College. He was in the prime of life and distinguished as a practitioner. His ardent temperament and undoubting self-reliance led him to the front in many controversial discussions; and his views were always maintained with force and ability. He was especially popular as a teacher; and his lecture hour is said to have been awaited by all with eager expectation. His sonorous voice and impressive manner, and the changing expression of his face, gestures and utterance, held the attention of his class and gave them an instructive entertainment rather than the didactic monotony of a lecture. In other directions he was equally active. He served as editor of the *American Medical and Philosophical Register;* he was for many years visiting physician to the New York Hospital; he was one of the founders and active members of the New York Historical Society; and his weekly evening receptions were the resort for all the social and literary celebrities of the day. He had an extensive acquaintance beyond the circle of his profession, and he attracted attention from all by the vigorous and enterprising qualities of his mind and disposition.

The professor of Surgery, *Dr. Valentine Mott*, who was then about thirty years of age and already rising into eminence, soon evinced a degree of capacity which made him the most celebrated operator in the United States. He was the pioneer in various exploits which were then uncertain ventures in the field of operative surgery, but which his carefulness and dexterity of manipulation almost invariably carried to success. His ligature of the arteria innominata, in 1818, is said to have made "an epoch in the life of the operator, and an era in the history of surgery." The ligature of the primitive iliac, the removal of the entire clavicle, and the resection of the inferior maxilla, extended still further his reputation for originality and skill. In the frequency and gravity of his operations he outranked all his contemporaries. He made the ligature of the common carotid artery forty-six times, and lithotomy more than a hundred and sixty times. In the opinion of his biographer he performed "a greater number of important and capital operations than any other surgeon who ever lived," and for more than half a century his reputation was "unequalled by that of any of his competitors in America, and scarcely surpassed by that of the most illustrious surgeons in Europe."

Not the least remarkable member of the faculty was *Dr. John W. Francis*, registrar and professor of materia medica. He had been the pupil of Dr. Hosack, with whom he was associated in business, and for whom he entertained a strong feeling of admiration. He was a devoted adherent of the new college organization, and served it to the best of his ability, both as registrar and professor. Beside his professional attainments his talent was largely in the direction of general literature; and both in his lectures and essays he was fond of "discursive utterances and amusing conceits." He delighted in historic reminiscences. His sketches and memoirs, though sometimes wanting in accuracy of detail, are full of entertainment; and his description of persons and things in the earlier half of the century forms a favorite volume in the publications of the New York Historical Society. He was among the founders and presidents of the New York Academy of Medicine, and was president of the medical board of Bellevue Hospital from the date of its organization.

Dr. Wright Post, appointed president of the College in 1822, was at that time fifty-six years of age. He had been professor for thirty years; first in the medical faculty of Columbia College and afterward in the College of Physicians and Surgeons. For nearly the same time he had been visiting surgeon to the New York Hospital; and he was everywhere esteemed for his skill as an operator, his fairness and delicacy in consultation, and his solicitude for the welfare of his patients. For twenty years he was "the undisputed head of the profession in this city;" attaining in no ordinary degree the reward of his labor in rank, wealth and reputation. In the estimation of his contemporaries his success was due to the general excellence of his mental and moral qualities, rather than to any marked endowment or peculiar gift. He had a refined and delicate physical organization, with a cool head, more judgment than imagination, and a manner which was always quiet and unobtrusive. In the lecture room his delivery was simple and natural; and he excelled in the presentation of truths attainable by patient industry. Notwithstanding the prominent position which he so long occupied, he seems to have been less deeply involved than many of his colleagues in the disputes and animosities of medical politics ; and he retained without interruption the friendship and respect even of those whose interests differed from his own.

Under these auspices the College became every year more widely known and more largely attended. A committee of the Regents reported it, in 1820, in a state of "rapid advancement." Its class in that year exceeded two hundred, being larger than at any former period; and in 1822, according to a report of the Regents, it had " an increased number of students from the most distant parts of the United States."

But notwithstanding this apparent prosperity there were at work causes of disturbance, which already impeded the operation of the College, and which soon grew to such proportions as to threaten it with disaster. The difficulty began, in 1819, with complaints from the Medical Society of the County of New York. After some correspondence, these complaints were laid before the State Medical Society and the Regents of the University. There were replies and countercharges from the professors, committees of investigation, and hearings before the Board of Regents; until the dispute culminated, in 1826, with the withdrawal of all the members of the faculty, and the appointment of new professors in their place.

At that time the element of personality entered largely into all discussions of a public nature; and there is evidence that it had its share in the college controversy. The members of the medical society, not directly connected with the College, were dissatisfied with its management by the faculty. They declared that this small body of men, whose talents and capacity they freely acknowledged, had formed among themselves a sort of "learned aristocracy" with its "train of favoritism;" and that they disregarded too plainly the claims and opinions of the profession. Under their direction the College had become a source of "exclusive privileges and immunities, to be exercised for their sole benefit;" and instead of serving, as it ought, for the general advancement of medicine in the city and State, it was used to promote the private interest of its possessors, and to supply honors and emoluments for their immediate friends and associates.

Beside the foregoing, more serious charges were brought against the faculty. The large classes in attendance, and especially the great number of graduates, instead of showing a wholesome growth of the College, were regarded as signs of its deterioration, due to a culpable laxity in the examinations. It was asserted that students had been graduated before attaining the age of twenty-one years; that some had received their diplomas after less than two years' study; that others had been graduated without ever having attended lectures in the College; and that it had even been announced in the circulars that no requisite for graduation would be expected, other than that of a satisfactory examination. This, it was said, was nothing less than an "invitation for full classes," intended to benefit the lecturers in the institution, at the expense of its character and usefulness.

Another cause of complaint was the large amount of fees exacted from students. In the earlier years of the College a definite and moderate sum had been fixed as the price for each lecture ticket, a liberal allowance being made for those who should take the entire course; and the president of each county medical society was invited to send to the College one student free of expense. But since then a different policy had been pursued. The lecture fees had been so increased as to make the expense of medical education greater than it ought to be; and the privilege of sending a free student, formerly accorded to the county medical societies, had been revoked. In some instances the vice of *pluralism* was said to exist in the professorial church; one professor holding two, or even three professorships at the same time, and requiring those who wished for either ticket to pay for all. Lastly, various additional fees had been established, making the total expense of a college course much greater than was originally intended.

All these evils were attributed to the control of the College being vested in the professors. They were subject, it is true, to the Regents of the University, whose consent was necessary to the adoption of by-laws and the granting of degrees; and who held the appointing power for both professors and trustees. But as professors and trustees were now practically the same, and formed the sole channel of official communication with the Regents, their interests and influence were unavoidably paramount, and excluded the governing body from other sources of information. They were, therefore, accountable for the alleged abuses in the management of the College, and for the public discredit into which it had unhappily fallen.

On the other hand, it was maintained by the professors that these attacks came from the jealousy of disappointed rivals, who had desired for themselves a share in the honors of the College and were envious of their more successful competitors. They claimed that the teachers in the several departments must be the most competent judges of a student's qualification for his degree; and that neither time of study, nor any other technical requirement, could supply a criterion equal to that of a professional examination. As to the expense of the college course, they denied that this was excessive; and the increasing classes in attendance showed that students did not consider it burdensome. To reduce the income of professors below the standard of respectability would not be for the interest of the College; which should be able to command the services of the best talent in the profession. They adduced a variety of reasons to justify the policy which they had pursued; and they protested against the assertions and inferences of their accusers as groundless or distorted exaggerations.

The views of the two parties as to their respective rights and duties were so divergent that no basis of accommodation could be reached, and the matter was referred to higher authority. A committee was appointed by the County Medical Society " to inquire into the condition of the College " and report thereon to the society. The report, which embodied the charges and criticisms above detailed, was followed by a memorial addressed to the Regents of the University, setting forth the causes of dissatisfaction, and praying that the College might be so reorganized as to exclude the professors from the Board of Trustees. It was also recommended that the new Board be composed of physicians resident in the city, together with the president of the College, the president of the County Medical Society, and the president of the State Medical Society; and that future vacancies in the Board be filled by the County Medical Society.

Early in 1820 this memorial was presented to the Regents. It was urged upon the attention of the Board with personal and documentary evidence, and was supported by a petition from many of the practitioners in New York. The professors were represented by a delegation from their own number, aided by the advice and co-operation of Mr. Emmet, the legal member of the Board of Trustees.

The Regents were by no means disposed to admit all the charges and recommendations of the medical society, and they deprecated the apparent feeling of hostility toward the College. But they recognized the existence of defects in its organization, and were convinced of the necessity for some remedial change. Especially the opinion advanced by the society, as to the bad effect of combining in a single body the functions of professors and trustees, was thought to be " well founded ; " and it was decided to separate them in future, so that trustees might no longer be " liable, merely from their situation as professors, to the suspicion of personal interest in the adoption of regulations." The professors who were already members of the Board were not displaced; but an ordinance was passed to the effect that " upon the death, resignation or removal from office of the present professors, or of any of them, no person or persons to be appointed as his or their successor or successors, shall be also eligible to the office of trustee; provided always that the president and vice president of the said College, for the time being, shall *ex-officiis* be trustees thereof." As the two officers above named were eligible from the faculty, and were usually selected from that body, this secured for the professors a virtual representation in the Board of Trustees.

The Regents furthermore fixed the lecture fees and other expenses of tuition at a definite rate, and required that each candidate for graduation must have studied medicine for three years with some respectable practitioner, and must have attended a full course of lectures at not less than two winter sessions. Lastly, they filled the existing vacancies in the Board of Trustees by the appointment of a number of resident practitioners, most of whom had been active in furthering the views of the medical society.

These measures, it was thought, would rectify the alleged faults in the administration of the College, and restore peace by removing the causes of complaint. But this proved a delusive expectation. The antagonism of individuals, already somewhat pronounced, was rather increased than diminished by bringing them in contact with each other; and it was soon apparent that the discord had only been transferred from the outside to the inside of the college organization. Previously, the contest was between the medical society and the professors; now it was between the professors and the trustees. The professors, who were no longer a majority of the Board, found themselves associated in council with their former opponents; and the newly appointed members believed that the needed reforms had been only half accomplished. Before long this division of sentiment became matter of public notoriety. It found expression in the printed notices and circulars of the College, and it was brought to the attention of the Regents in repeated memorials, reports and protests from both sides. The Regents, instead of being discouraged by the apparent failure of their efforts, were again ready to give the subject patient consideration. It is evident, from their proceedings, that they felt a deep interest in the welfare of the College, and spared no pains to compose its difficulties. In April, 1825, they appointed a committee "to visit the College of Physicians and Surgeons in the city of New York, to the end that the matters in controversy between the trustees and professors thereof, and the affairs of said College generally, may be more fully investigated."

The members of this committee performed their duty with unwearied fidelity. They visited the College in the summer vacation, and invited the trustees and professors to meet with them for a free interchange of views, continuing the sessions daily for an entire week; and they embodied the result in a report to the Regents covering fifty-four manuscript pages.

The report declares that "it is due to the professors and trustees to say that, however much they differ from each other in the arguments urged or in the conclusions drawn from the facts inquired after, yet in the whole course of the business the conduct of each was characterized by an openness of communication and an integrity of manner which gave assurance of the fairness of the motives of each; and that their conclusions, maintained oftentimes with vehement warmth, were urged from a conscientious belief in their correctness, and that they were essential to the welfare and advancement of the College." The committee found the most damaging of the accusations to be unsustained; and they relieved the professors from any serious imputation on their personal integrity. The existence of either partiality or oppression in the examination of candidates appeared to be an improbable inference ; and there was no suspicion of misapplication or maladministration of the college funds. But the controversy had nevertheless destroyed that " union of sentiment and concert of action " which ought to exist between professors and trustees; and it had assumed a form which would hardly admit of cordial reconciliation. It appears that, in the minds of the committee, this unfortunate discordance was due in great measure to the existence of " professional rivalries ; " and they were of opinion that it could only be cured by so changing the membership of the Board of Trustees that it should no longer consist wholly of medical practitioners.

The Regents adopted this recommendation; and at their next meeting an ordinance was passed in the following terms :

"Be it therefore ordained by the Regents of the University of the State of New York, that all vacancies now existing, or which shall hereafter happen, in the board of trustees of the College of Physicians and Surgeons in the city of New York, shall be filled by the appointment thereto of persons who are not of the medical profession; until the number of such persons, so to be appointed, shall constitute at least thirteen of said board."

This provision has remained a fundamental part of the college organization from January, 1826, to the present day. The wisdom of its framers has been justified by the experience of many years; during which time no serious effervescence has occurred, like that which formerly endangered the prosperity of the institution. The government of the College preserves its professional character through the medical members of the Board of Trustees; and it has the aid and counsel of capable men in other walks of life, who can give to the College the benefit of their varied information, and who are free from participation or interest in the rivalries of medical ambition. For more than half a century the practical operation of this plan has answered all reasonable expectations.

Nevertheless, the pacification of the College was not accomplished without a further struggle. The faculty at this time was considerably smaller than when it received the accession of professors from Columbia College. Resignations, removals, and deaths had reduced its number to six; but these embraced the most important and influential of its members. In the course of the controversy they had demanded, as an essential measure for the preservation of the College, the entire renovation of the Board of Trustees, by the removal of all its members and their replacement by non-medical men. They were not satisfied with the gradual and partial change adopted by the Regents; and they intimated plainly that they could not retain their positions unless it were made complete. But the committee of investigation had acquitted the trustees, as well as the professors, of culpable malfeasance; and reported that, in their opinion, "if the Regents possess the power, it ought not to be exercised to remove all or any of the trustees." The professors still hoped to accomplish their object by carrying it before the legislature. But this also proved a failure; and when the question was finally decided against them, they sent to the Board of Regents their formal resignations, April 11th, 1826.

This resignation, however, was by no means a retirement from the field; it was only a change of base for future operations. The men who had been prominent and successful teachers for so many years could not be expected to surrender their occupation as well as their appointments. They believed that they had given to the College as much as the College had conferred upon them; and that they might carry with them elsewhere the prestige of their undisputed ability. They determined to organize at once a new medical school, which should be the rival of the old one, and in which they could exercise their functions as professors, unimpeded by the control of a hostile authority. For this purpose they obtained the assent of Rutgers College, in New Jersey, to the formation of a medical department officially connected with that institution. They fitted up, at their own expense, a building in Duane street; where, in the autumn of the same year, they inaugurated their course of instruction under the name of the *Rutgers Medical College*.

But the new enterprise, undertaken with such persistent energy, failed to meet with the desired success. The resignation of professors from a medical school, whatever their superiority of talent, does not always result in its injury or downfall. It is apt to turn out that there are others who are able to fill the chairs they have left, and who can still maintain the credit and prosperity of the institution. That is what happened in the present case. The Regents of the University, at a special meeting in July, 1826, appointed for the College of Physicians and Surgeons a new corps of professors, who long held their places with distinguished honor, and of whom three became afterward presidents of the institution. The Rutgers Medical College had an existence of only four annual sessions.

CHAPTER III.

REORGANIZATION OF THE COLLEGE.

ITS DIFFICULTIES, PERSEVERANCE AND SUCCESS.

1826–1837.

The constitution of the faculty under the new organization was as follows :

THE FACULTY IN 1826.

JOHN WATTS, M.D., *President.*
JOHN AUGUSTINE SMITH, M.D., *Professor of Anatomy and Physiology.*
JAMES F. DANA, M.D., *Professor of Chemistry.*
JOHN B. BECK, M.D., *Professor of Botany and Materia Medica.*
ALEXANDER H. STEVENS, M.D., *Professor of the Principles and Practice of Surgery.*
EDWARD DELAFIELD, M.D., *Professor of Obstetrics and the Diseases of Women and Children.*
JOSEPH M. SMITH, M.D., *Professor of the Theory and Practice of Physic and Clinical Medicine.*

The fourth president of the College, DR. JOHN WATTS, was a practitioner of eminence, visiting physician to the New York Hospital, and consulting physician to the New York Dispensary and the Lying-in Asylum. At the time of assuming the presidency he was forty years of age, and in the full exercise of his professional activity. He was already a member of the Board of Trustees, to which he had been appointed in 1820, as one of the representatives of the County Medical Society in the management of the College. He was thus personally familiar with the recent history of the institution; and his general reputation for sagacity and integrity, as well as his acknowledged professional attainments, enabled him to contribute in great measure to the restoration of its prosperity. He continued to serve as presiding officer until his death in 1831.

Of the professors appointed in 1826, several had already some experience in teaching. Dr. John Augustin Smith, formerly the professor of anatomy and surgery, had resigned his chair in 1814 and removed to Virginia, where he became the president of William and Mary College, his alma mater. On the reorganization of the College of Physicians and Surgeons he returned to New York and resumed his connection with the institution, as professor of anatomy and physiology. Dr. Dana, the professor of chemistry, had served with credit for several years in the same capacity in Dartmouth College, New Hampshire. Dr. Stevens had been for nearly ten years visiting surgeon to the New York Hospital, where he was well known for his success as a clinical lecturer, both at the bedside and in the amphitheatre; and his private office was frequented by an unusually large number of pupils. Dr. Delafield was surgeon to the New York Eye Infirmary, which had been for several years in operation. Dr. Joseph M. Smith had attracted attention as a writer and reviewer; and Dr. Beck was chief editor of the *New York Medical and Physical Journal*, then in the fifth year of its existence. Three of the faculty were graduates of the College; and with the exception of Dr. Watts and Dr. John Augustine Smith, none of them had reached the age of forty years. But they were by no means untried; and the promises which they had already given, for character and ability, were afterward abundantly realized.

For some years subsequent to its reorganization the College was beset with continuous and harassing difficulties. It had experienced a revolution which restored its internal harmony, but at dangerous cost of time and means. There were still many obstacles in the way of its success. Its resources were at a very low ebb. In August, 1826, a committee was appointed by the Trustees to make an inventory of the college property, to consider all matters relating to the interests of the institution, and to devise measures for its relief. It appears from their report that the property of the College then consisted of little more than its land and building in Barclay street, and that there were against it outstanding claims, mainly from the former professors, covering nearly the whole of this value; so that if these claims were to be allowed and satisfied, even without a forced sale of the property, there would remain in the hands of the College "not more than $4,570." The building, moreover, was in bad condition. The committee found the "seats and furniture of the lecture rooms very much out of repair;" the anatomical museum "almost empty;" the "shelves and cases of the professor of chemistry empty;" some of the fixtures belonging to the same department "taken down and carried away;" and the College generally presenting the appearance of a "city sacked and deserted by a ruthless enemy." There was left in its possession but a slender remnant of material and apparatus for teaching purposes; a deficiency which could only be supplied by the voluntary exertion of its friends.

Beside these sources of anxiety in the internal affairs of the College, there were others equally threatening without. The retirement of the former faculty, and the organization of a new medical school under their direction, not only created an earnest and able rivalry against the college; it also established a strong partisan opposition between the adherents of the two institutions; and this extended so far as to enlist, in favor of one and against the other, nearly every medical man in the city and not a few in other parts of the State. The traces of this feeling, in the records of the College and in the medical literature of the day, are sufficient to show that it gave rise to much difference of opinion and hostile criticism on both sides.

But the new professors were equal to the emergency. They made every effort to provide for the wants of the College and to restore its efficiency. The important anatomical collection belonging to the former professor, which had been sold and removed from the building, was purchased by a friend of the institution, and by a "liberal and spirited arrangement" between the purchaser and the professors was returned to the museum of the College. Its shelves were also replenished from the private collections of Dr. Stevens and Dr. Delafield, with further contributions from other similar sources. The New York Eye Infirmary was thrown open for clinical instruction; and a special course of lectures on diseases of the eye was given by Dr. Delafield, in addition to those of his obstetrical chair. By persevering attention to the duties of their professorships, the members of the faculty demonstrated their ability to maintain their position as teachers, and to meet with success the competition of their elders and predecessors.

At the first session of the College after its reorganization the number of students in attendance was reduced to ninety. But it soon began to increase, and in five or six years it again reached nearly its former standard. The "Rutgers Medical College," established at first under the authority of Rutgers College, New Jersey, afterward transferred its allegiance to Geneva College, in the State of New York. But it failed to obtain, under either of these connections, the requisite legal sanction for its diplomas; and the attempt was finally abandoned. About the same time the pecuniary claims against the College of Physicians and Surgeons, which had been pressed with much urgency by suits at law, were adjudged and settled on a more favorable basis than had been anticipated.* The College was thus relieved from the most vexatious of its financial embarrassments, and freed from the annoyance of professional opposition. Its vitality had been tested and strengthened by the trials it endured, and it gained at last the permanent respect of its opponents as well as its friends.

All this time a strict economy was required and practised, to maintain the safety and independence of the institution. The unexpired lease of the premises on Park Place, in the rear of the College, was sold; the College retaining for its accommodation only the "conveniences for a sink." All subscriptions to foreign and domestic periodicals, and all purchases of books, were discontinued. Various perquisites and commissions, formerly allowed to the registrar, treasurer, librarian and president, were abolished; these officers thereafter performing their duties gratuitously. The graduation fees, previously divided among the faculty, were assigned to the general income of the College; and for two years a tenth part of the lecture fees of each professor was appropriated to a fund for reducing the indebtedness of the institution. By these means the College was gradually placed in a position of comparative security, and enabled to provide for its most important liabilities.

* These claims were originally for $20,478. They were settled, according to a judgment rendered in 1830, for $13,986.

Among the active members of the Board of Trustees, who aided in sustaining the College during this period of contest and difficulty, were *Dr. John Kearny Rodgers*, the eminent surgeon and accomplished anatomist, who was visiting surgeon to both the New York Hospital and the New York Eye Infirmary; *Dr. John D. Jaques*, the faithful and efficient treasurer of the College since 1811, who continued to perform the duties of that important office, without compensation, till 1838; Dr. Nicoll H. Dering, the assiduous registrar, a graduate of the class of 1817; *Drs. William Hamersley* and *Francis U. Johnston*, both physicians to the New York Hospital; *Drs. Samuel Borrowe* and *John C. Cheesman*, surgeons to the same institution; *Mr. James A. Hamilton*, a son of Alexander Hamilton; the Hon. *Elisha W. King;* the Hon. *George W. Bruen;* the Hon. *James Campbell;* the Hon. *Stephen Allen*, who had been Mayor of the city in 1821, 1822 and 1823; *Mr. Charles G. Troup;* and *Mr. Samuel Boyd.* Of the non-medical trustees above mentioned, two were merchants and the rest members of the legal profession.

Meanwhile, some changes occurred in the faculty. The first was due to the death of Dr. Dana, the professor of chemistry, in 1827, within a year after his appointment. He was considered a man of great promise, and was universally regretted by his pupils and his colleagues. His successor in the chemical chair was Dr. *John Torrey*, a graduate of the class of 1818, who had been for three years professor of chemistry, geology and mineralogy in the United States Military Academy at West Point. He at once entered on his duties in the College of Physicians and Surgeons, where he remained for many years, acquiring a wide reputation both as chemist and botanist, and contributing much to the scientific repute of the institution. In compliance with the express desire of himself and Dr. Beck, the department of botany was transferred to Dr. Torrey and that of medical jurisprudence to Dr. Beck. By this exchange each professor was placed in charge of the subject most adapted to his taste and acquirements.

Subsequently, *Dr. Valentine Mott* again became professor in the College. His increasing reputation and success as an operator made him a desirable acquisition for the corps of teachers; and on his part the disposition to rejoin the College was not wanting. For the purpose of carrying out this design a new chair was established in 1831, to which Dr. Mott was appointed, as professor of "Operative Surgery and Surgical and Pathological Anatomy." At the solicitation of the faculty he was also associated with Dr. Stevens in the clinical teaching of surgery at the New York Hospital. He continued to occupy this position for nearly four years, when he was compelled to suspend professional work and to undertake a European tour for the benefit of his health. Owing to his prolonged absence, the chair which had been created for him was abolished in 1837; and after his return he received the complimentary appointment of Professor Emeritus.

In 1834 the chair of anatomy and physiology, held by Dr. John Augustine Smith, was divided; Dr. Smith remaining professor of physiology, and *Dr. John R. Rhinelander*, a graduate of the class of 1824, becoming professor of anatomy. He had already served in the College for five years as Demonstrator of anatomy; and on assuming the professorship he deposited in the college museum a large collection of specimens prepared by himself.

On the death of the president, Dr. John Watts, in 1831, his place was filled by the appointment of *Dr. John Augustine Smith*. The vice president was Dr. Thomas Cock, who had been the incumbent of that office since 1827.

The next event of importance was in 1837, when the College was removed to Crosby street, about one mile farther up town. For some years the old building had been very inadequate in its accommodations; beside which the movement of population away from its vicinity, and the increase of business establishments near by, made it evident that the College must soon need a more appropriate location. The matter was discussed in the Board of Trustees as early as 1834; and in the following year a committee was appointed to "ascertain whether a new site could be obtained for the College and upon what terms," as well as to "procure plans and estimates of the expense attending the erection of proper buildings." Soon afterward the committee reported that a piece of property in Crosby street, belonging to the *New York High School Society*, was for sale; and that, with certain alterations, the building might be made serviceable for the purposes of the College. The lots were on the east side of the street, having a frontage of seventy-two feet and a depth of one hundred feet; the building being of considerably smaller dimensions. In accordance with the recommendation of the committee, the property was purchased, in May, 1835, for $18,500; but it was subject to an unexpired lease, and consequently did not come into the possession of the College until two years later. As early as possible thereafter plans for the reconstruction of the building were submitted to the Board of Trustees and approved by them. The work was commenced on the first of June and completed by the first of November; and the building was inaugurated at the opening of the thirty-first session, November 6th, 1837.

The removal of the College was accomplished at a time very unfavorable for such an enterprise. The disastrous fire in the winter of 1835–36, which destroyed thirteen acres of the most important business structures in the lower part of the city and caused a general collapse of the insurance companies, was followed by the scarcity and distress of the succeeding winter, and lastly by the financial revulsion of 1837, when all the banks in the United States suspended payment for a year. These misfortunes impeded, for the time being, every kind of business operation. It was difficult to find a purchaser for the college property in Barclay street; and when finally sold, it brought only $31,000. Of this sum, nearly one half was covered by a mortgage which had been placed on the property, to meet the judgment rendered against the institution in 1830; and there had been, for the past two years, a diminution in the receipts of the College, owing to the general disturbance and uncertainty in financial affairs.

The entire cost of reconstructing and furnishing the new building was $14,370. But the funds remaining in the hands of the trustees, after paying for the original property, amounted to less than $10,000. This left a deficit of $4,450, which had to be supplied by a loan from the individual professors; the largest share of this financial burden being borne by Dr. John Augustine Smith. All things considered, it appeared to the trustees a matter of congratulation that "notwithstanding the pressure of the times" the work had been successfully completed according to the original design.

The new building was regarded as a great improvement on the former structure; "imposing in its appearance, ample in its dimensions, and commodious in its arrangements." In the last respect especially it was confidently declared to be "unsurpassed by any similar establishment in the Union."

The edifice was of brick, three stories high, with a lecture room on each floor; sixty-five feet wide in front and seventy-five feet in depth, with an open passage way from front to rear along the north side. The first floor, nearly level with the street, contained the chemical lecture room, calculated for an audience of three hundred, with the professor's laboratory and apparatus rooms in the rear. On the second floor

THE COLLEGE BUILDING IN CROSBY STREET.
1837–1856.
From the College Circular for 1837.

was the College Hall, or principal lecture room, used also for Introductory and Commencement exercises; a cabinet of Materia Medica; and private rooms. The third floor was occupied by the amphitheatre, the museum, private rooms, and the general dissecting room. The amphitheatre, the lecture rooms, and the dissecting room were warmed by stoves, the smaller apartments by open fire places. All were lighted with gas and supplied, after a few years, with Croton water; neither of which conveniences had existed in the Barclay street building. In 1847 an additional story was constructed on the rear, giving a larger dissecting room and more space on the third floor for other purposes. The building was known as No. 67 Crosby street, and was occupied by the College for a little over eighteen years.

CHAPTER IV.

THE COLLEGE IN CROSBY STREET.

ERA OF IMPROVEMENT.

1837-1856.

About the time of the removal of the College to Crosby street the faculty received a number of important accessions. In 1837 Dr. Stevens resigned the professorship of surgery, which he had held with distinguished ability for eleven years, and was followed by Dr. Alban G. Smith, previously professor of surgery in the Medical College of Ohio. Dr. Smith however lectured for only two sessions. The chair was then filled by *Dr. Willard Parker*, who had been for four years professor of surgery in the Medical College of Cincinnati. On receiving his appointment in 1840, Dr. Parker removed to New York, where he at once gave evidence of superior talent, both as professor and practitioner.

In 1839, the chair of anatomy having been vacated by the resignation of Dr. Rhinelander, *Dr. Robert Watts* was made professor in that department. He was a nephew of the former president of the College, Dr. John Watts, and a graduate in the class of 1836.

At the time of his appointment he was twenty-seven years of age, and had already taught anatomy in the medical schools of Pittsfield, Massachusetts, and Woodstock, Vermont. His untiring industry and accurate information made him a valuable addition to the faculty and a marked favorite with the students of the College.

Nearly at the same time Dr. Delafield retired from the chair of obstetrics, and in 1841 this professorship was filled by the appointment of *Dr. Chandler R. Gilman*, a well-known practitioner in the city, who had officiated as lecturer on obstetrics for the preceding session.

The fifth president of the College, DR. JOHN AUGUSTINE SMITH, is an interesting character in the history of the institution from his long and varied connection with its affairs. He was identified with its original organization, having taken part in the first session as lecturer on anatomy, and receiving immediately afterward the appointment of professor of anatomy and surgery. He continued in this chair until 1814, when he removed to Virginia, his former home, to accept the presidency of William and Mary College in that State; returning after an interval of twelve years, to resume his connection with the College of Physicians and Surgeons, in 1826, as professor of anatomy and physiology. After becoming president in 1831, he continued to lecture on anatomy and physiology until the chair was divided in 1834, and subsequently on physiology until 1843; thus combining for a period of twelve years the duties of president and professor.

His character was marked by an amiability of disposition and a courtesy of manner which survived all the trials and difficulties of his administration. These qualities were especially noticeable in his relations with Dr. Wright Post, who was attached to the chair of anatomy in 1814 as joint professor; Dr. Smith leaving the College soon afterward, and in turn replacing Dr. Post at its reorganization in 1826. Notwithstanding these professional complications, Dr. Smith was selected by the County Medical Society to prepare the obituary memoir of Dr. Post in 1828. In this address he refers to their intercourse as follows:

"A good understanding always existed between Dr. Post and myself. At first in opposition, then joint professors, and lastly independent friends; no hostile feeling was at any time manifested with regard to each other, nor did any, I am persuaded, ever exist. And now it has fallen to my lot to hold up, however feebly, to the just admiration of his fellow citizens the only man with whom, in the whole course of my life, I have come into any sort of collision, whose talents and station could for a moment induce me to consider him as a rival."

Dr. Smith was equally good-tempered in replying to criticisms on himself. He was thought by some to be more versed in philosophic generalities than in the exact details of specific information. Nevertheless, in an introductory lecture delivered in 1827 he expressed a strong disapproval of "medical theorists;" and when charged with inconsistency in this respect, he replied as follows:

"With these remarks the author takes leave of medical theorists; adding however that his friends have suggested to him that he himself, from his propensity to generalize, is obnoxious to his own censures. If such be the fact, it only adds one to the long list of those whose precepts are better than their example; and, he sincerely hopes that his brethren will profit by what is right, while, more fortunate than himself, they eschew what is wrong."

Dr. Smith was a man of scholarly attainments, somewhat given to metaphysical discussion, but a polished writer and an agreeable and instructive lecturer. Though not distinguished for any remarkable traits of scientific originality, he seems to have had, much influence with his associates, and to have conducted the affairs of the College with judgment and success. He resigned the presidency in 1843, when Dr. Stevens was appointed in his place. At that time the department of physiology was again attached to the chair of anatomy.

In consequence of these changes the faculty of the College, soon after its establishment in Crosby street, became virtually a new one, with the following organization:

THE FACULTY IN 1843.

ALEXANDER H. STEVENS, M.D., *President.*

JOSEPH M. SMITH, M.D., *Professor of the Theory and Practice of Medicine and Clinical Medicine.*

JOHN B. BECK, M.D., *Professor of Materia Medica and Medical Jurisprudence.*

JOHN TORREY, M.D., *Professor of Chemistry and Botany.*

ROBERT WATTS, M.D., *Professor of Anatomy and Physiology.*

WILLARD PARKER, M.D., *Professor of the Principles and Practice of Surgery.*

CHANDLER R. GILMAN, M.D., *Professor of Obstetrics and the Diseases of Women and Children.*

The period from 1837 to 1856, during which the College remained in Crosby street, was one of substantial progress in reputation and prosperity. Its average attendance of students increased from one hundred and forty, as in the preceding ten years, to about two hundred. The traces of antagonism in various quarters, the legacy of its former turmoils and dissensions, disappeared before the growing popularity of its teachers and the united support of its officers and trustees. This opened for the College a new prospect and placed it in a different position. Heretofore its energies had been consumed in an unavoidable conflict with difficulties. Now they were employed to enlarge its resources and increase its usefulness.

An important innovation introduced at this time was a change in the mode of *selecting and appointing professors.* In 1837 the plan was inaugurated of providing for a vacant professorship by the appointment of a " Lecturer," who should perform the duties of the chair during one session before he could be eligible as professor. From the earlier years of the College it had been sometimes the practice for the Trustees to appoint lecturers; but this was only as a temporary expedient, to fill a vacancy while awaiting the action of the Regents, or to provide for lectures during the absence or disability of a professor. It was now adopted as a rule, that no one should be appointed to a professorship until he had served as lecturer to the satisfaction of his colleagues. Every candidate was thus put to a preliminary test; and the College was relieved from the danger of hasty or injudicious appointments. After a few years' trial the advantages of this method were so obvious that it received the approval of all and became the established policy of the institution. It has been retained without interruption to the present day.

Furthermore, the instruction given in the regular term was supplemented by courses of *Spring and Fall lectures.* Soon after the organization of the College there had been a course of lectures in April, May, and June, devoted to zoology, botany, mineralogy, and chemistry. But this subsequently fell into disuse. At first, two of the professors took part in it, afterward only one; and since 1817 it had been wholly discontinued. It was now revived, on a larger scale and with a different object; namely, to give instruction on special subjects, mainly of a practical nature, which could not be sufficiently treated during the regular session. In 1841 a Spring course was inaugurated, with two or three lectures a day from March till June. It was conducted by a corps of eight lecturers, including professors Parker, Watts, and Gilman, as representing the faculty; Dr. Quackenboss, the demonstrator of anatomy; and four other physicians connected with various medical institutions in the city. The programme for this course was as follows:

SPRING COURSE OF LECTURES FOR 1841.

Operative Surgery.......Dr. WILLARD PARKER.
Surgical Anatomy.......Dr. ROBERT WATTS.
Pathology of the Uterus..Dr. CHANDLER R. GILMAN.
The Nervous System.....Dr. JAMES QUACKENBOSS.
Diseases of the Eye......Dr. GEORGE WILKES.
Pathology of the Thorax .Dr. ALONZO CLARK.
Club feet and kindred deformities...........Dr. WILLIAM DETMOLD.
Pathology of the Kidney..Dr. WILLIAM C. ROBERTS.

In the Fall of the same year a "preliminary course" was established, continuing through October to the opening of the regular term in November. This course was conducted wholly by members of the faculty, and comprised the following subjects:

PRELIMINARY FALL COURSE FOR 1841.

Medical Botany	by	Dr. Torrey.
Comparative Osteology	"	Dr. Watts.
Pathology of the Ear	"	Dr. Parker.
Monstrosities	"	Dr. Gilman.

The Spring and Fall lectures were soon found to be valuable additions to the regular course. They were not obligatory upon the students as a requisite for graduation; but to all who could avail themselves of the privilege they afforded increased opportunities for acquiring information. They were continued and assiduously sustained by the faculty. The announcement for 1844 states that the Fall course is to be enlarged, to include lectures by "all the professors," making the entire session "virtually of five months' duration;" the customary period of four months, required by law, being, in the opinion of the faculty, "too short even for the regular course, and much too short to allow them to enter into specialities."

Three years later the regular course was extended to four months and a half, beginning in the middle of October; and the preliminary course was placed at an earlier date, beginning in September.

The Spring course was also enlarged by securing the coöperation of various medical men who were known as experts on special subjects; becoming in this way a "valuable and practical course on specialities." Beside the topics already enumerated, it included, in the following years, such subjects as Poisons, Diseases of the Skin, Comparative Anatomy, Pathology of the Intestines, Dislocations and Fractures, Uterine Hemorrhages, the Physiology of Generation, Pathology of the Urine, and Diseases of Infancy. These lectures were appreciated by graduates as well as by students; and they were also useful by bringing into notice the acquirements and talent of younger members of the profession in various departments of practical or scientific medicine.

At the same time the faculty began to pay greater attention than before to *material illustration* as a means of instruction. In the announcement for 1837, especial stress is laid on the increased facilities in the new building for practical anatomy, and on those for illustration in various departments by specimens, drawings, models, wax preparations, and plaster casts. The anatomical specimens belonging to the former professor in this department were purchased and secured for the benefit of the College. From time to time there were other contributions by former graduates. In 1845 the museum was enriched by the collection of Dr. J. Kearny Rodgers, an old friend and trustee of the institution; and in the same year professor John B. Beck presented his cabinet of materia medica, containing nearly six hundred specimens. On all sides a desire was manifested to enlarge the means of instruction beyond those of a strictly didactic course. The announcement for 1850 declares that "the great object of the faculty is to make the courses of instruction as demonstrative and practical as possible; and in this object they are warmly seconded by the Trustees of the College."

One of the most important features of this policy was the *College Clinic*, established in 1841 by the sagacity and enterprise of Dr. Willard Parker, then recently appointed professor of surgery. In the preceding year a number of private pupils had been taken to the Northern Dispensary, at the intersection of Christopher street and Waverley Place, to witness there the methods of diagnosis and treatment. This was found so useful that soon after the commencement of the session it was "thought best to make the College the place for this kind of instruction," so that all might share in its benefits. Outdoor patients were accordingly brought, from the Dispensary and elsewhere, to the College building, to be examined and treated in the presence of the class.

This was the beginning of the entire system of college clinics, which have since grown into such magnitude. At first they were held once a week during the session; but they were soon increased in frequency and continued through the intervening months. In the catalogue for 1843-4 it is announced that "a Clinique has been established in the College, which has been in successful operation for several years, under the direction of the professor of surgery. The cases presented to the class embrace almost every variety of minor surgery, and many of them require operations which are performed before the class."

. . "In the course of the present session the professor of obstetrics has made arrangements for a clinique to be held every Thursday, and to embrace the diseases of women and children."

Under this plan the number of patients rapidly increased, and the clinic grew in favor with both the students and the profession. In 1844 it is styled the "Medical and Surgical Clinique," and the encouragement received in its support is duly acknowledged. "The great number of Medical men in New York, many of them connected with public institutions and desirous of promoting the interest of the College, furnishes the Clinique with an abundant supply of useful and interesting cases, embracing Medical and Surgical diseases; also those of Children. These cases are examined and prescribed for before the class, so that each student can follow for himself the causes, symptoms, diagnosis, prognosis, and treatment of every case." The scope of the enterprise continued to enlarge; and in 1850 it had "assumed a degree of importance that could hardly have been anticipated at its origin." Subsequently there were three clinics each week at the college building, throughout the regular session and during the Spring and Fall terms.

The advance of the College at this time, in efficiency and repute, was largely due to the unremitting exertion of its professors, and their judicious employment of various means for its improvement. They felt that the course of instruction needed a greater expenditure of time and attention on their part than had yet been given to it; and they were ready to adopt new methods and measures whenever occasion required. The master spirits in this movement were Drs. Parker, Watts and Gilman. They were ambitious in the pursuit of professional distinction, but with no wish to revive the personal differences of former times, now happily becoming obsolete; and they knew how to enlist the sympathy and coöperation of others in an institution intended for the benefit of all. The College will always be indebted to their energy and foresight for inaugurating improvements of lasting influence on its future welfare.

Beside these changes in the internal affairs of the College, there were others, connected with the general growth of the city and its institutions, which aided the advancement of medical education. One of the most important was the reorganization of *Bellevue Hospital*. This institution was originally the city almshouse, which in 1807 was located in Chambers street. At that time it had no regular hospital organization; but such of its inmates as required medical treatment were cared for by an attending physician at his stated visits. In 1816, owing to the need of larger accommodations for the poor, a new building was erected on the shore of the East River, at a place known as "Bellevue." This was formerly a country seat belonging to Lindley Murray, the grammarian, who describes it, in his Letters, as "most delightfully situated," commanding an extensive view up and down the river, with a fruit and flower garden, and a field in the rear for pasturing cattle. After the erection on this ground of the almshouse proper, one or two smaller structures were added near by, for the reception of the sick. But with the growing demand on the city charities, these accommodations also proved insufficient; and in 1848 the almshouse department was removed to Blackwell's Island, leaving the entire establishment at Bellevue for use as a hospital.

In this way Bellevue hospital was added to the medical institutions of the city. But in the following year it was greatly improved through a reorganization of its management under the authority of the State legislature. By this law its affairs were confided to a Board of Governors, who, acting in concert with the medical profession, introduced into its administration several much needed reforms. One of the improvements so effected was the opening of the hospital for clinical teaching. It had now become a much larger establishment than before, admitting over three thousand patients annually; and it would consequently afford, if accessible to students, much valuable opportunity for medical instruction.

This was accomplished during the session of 1849-50, through the influence of the attending physicians and surgeons with the Board of Governors. In the "Rules and Regulations" of the Board, published in 1851, the admission of students was authorized in the following terms:

"In order to render the hospital, as far as it may

consist with the welfare of the patients, conducive to the advancement of medical science, the physicians and surgeons may provide among themselves adequate and regular practical instruction, by observations accompanying operations, by clinical lectures or otherwise, to the students admitted to see the practice of the house, during the ordinary periods of lectures at the medical institutions of the city, and longer if deemed expedient."

The plan was immediately carried into effect and its advantages appreciated. In the annual report of the Medical Board of the hospital for 1851 it is said that "clinical lectures have been given regularly at the hospital during the fall and winter months, and numerous surgical operations performed, which have been attended by classes of medical students numbering from fifty to two hundred and fifty." Medical instruction was thus more amply provided for than at any former period. The college clinics exhibited the early stages or milder forms of disease; while cases of graver character or more advanced condition were to be seen at the New York and Bellevue hospitals. In both these institutions the College was represented; at the New York hospital by Dr. Joseph M. Smith, attending physician, and at Bellevue by Dr. Parker, attending surgeon, and Drs. Gilman and Clark, attending physicians.

Another noteworthy event was the *legalization of practical anatomy* by act of the legislature in 1854. This had long been an object of earnest desire with all interested in medical education. The only sources of supply for anatomical material under existing laws, in the State of New York, were the unclaimed bodies of convicts dying in the State prisons at Sing Sing and Auburn; and these were entirely inadequate for the purposes required. In 1843 and 1844 attempts were made to remedy the evil by the introduction of a bill authorizing the use of unclaimed bodies from all the public penal and charitable institutions of the State. The county medical societies of Erie and Onondaga officially addressed the State Medical Society, asking its coöperation in favor of the proposed law. But the committee of the State Society, to which the matter was referred, reported that "in view of the present state of the public mind," it would be inexpedient to take action at that time; and the subject was accordingly dropped. It was taken up afresh in 1853, when a similar bill was framed and advocated by a few medical men who were members of the legislature. Dr. Alonzo Clark, then president of the State Medical Society, in his anniversary address delivered before the society and the legislature in joint session, strongly urged its adoption, using every effort to demonstrate its propriety and to dissipate the prejudice against it. He was so nearly successful that the bill passed the Assembly, but was unexpectedly lost in the Senate.

In the following year the attempt was again renewed. The bill was placed in the hands of the Hon. Frederick A. Conkling, who introduced and supported it in the legislature with much earnestness and discretion. Dr. Parker, with other medical men from New York, spent much time at Albany, explaining its object and justifying its provisions to the members of the legislature. At the instance of the Queen's county medical society, the State society adopted resolutions approving the bill and recommending its passage; and Dr. Sprague, the president of the State society for that year, made a strong appeal in its behalf. These combined efforts at last effected the object; and the bill entitled "An Act to promote Medical Science" was passed April 1st, 1854.

The good effect of this measure was at once apparent. Teachers of anatomy were enabled to make their instruction demonstrative and practical, without being compelled to rely on the uncertain and illegal aid of resurrectionists. The college catalogue for 1855 announces that the preceding session has been "distinguished by a new element of success." . . . "Thanks to the enlightened liberality of the legislature who passed the *Anatomical Bill*, the supply of subjects has not only been ample, but it has been obtained without the difficulties and dangers of former years." At that time Massachusetts was the only state already provided with a general anatomical law of this kind. Since then enactments similar or equivalent to that of New York have been adopted in twenty-one other states of the Union.

The progress of the profession during this period was also marked by the formation of the pathological society and the academy of medicine. The *Patholog-*

ical Society originated in 1844 by the association of twenty-three medical men, including several of the professors and lecturers in the College. For some months the society met at the offices of its various members; but after the first year its meetings were held in the anatomical theatre of the College, where greater facilities were afforded for the exhibition of specimens and the accommodation of the audience. At the end of two years its original membership had more than doubled; and since then it has continued to grow in prosperity, and in the interest and usefulness of its proceedings. In 1886 it received an endowment of Five Thousand dollars from Dr. Middleton Goldsmith, of Rutland, Vermont, one of its original members and a graduate of the College in the class of 1840, to establish, under the auspices of the society, periodical lectures on subjects connected with pathology. The society now numbers over two hundred members.

The New York *Academy of Medicine* was organized in 1847 for the cultivation of medical knowledge by oral and written communications, and for maintaining, by precept and example, an elevated tone of professional feeling. It was intended to "represent if not embrace" the great mass of regular practitioners in the city. Its first president was Dr. John Stearns, a graduate and former trustee of the College; its treasurer was professor Watts; and upon its standing committees were Drs. Alexander H. Stevens, Edward Delafield, John B. Beck, Willard Parker, and Joseph

M. Smith. Almost immediately the academy became the recognized place of discussion for prominent questions of medical interest, and for the presentation of new views, theories or observations. It added much to the opportunities for mutual criticism and improvement, and opened for its members a new and useful field of professional activity.

In all these movements to promote the interests of the profession, the College had an important share. Its officers and teachers were among the most active in furthering every design for the advancement ot medical knowledge or the welfare of medical practitioners; and it was always an advocate for means of improvement and progress. It was beginning to repay, after many years, the earnest devotion of its founders, and the persevering efforts of its faculty, friends and graduates. It was aided in its growth by the increased facilities for medical instruction in the city at large, to which it contributed in great measure by its own exertions and influence.

CHAPTER V.

CHANGES IN THE FACULTY.

1851–1858.

Owing to changes by death and resignation during the stay of the College in Crosby street, the faculty lost some of its most prominent members; and the arrangement of professorships was somewhat modified, in order to supply the vacant places.

As early as 1842 the professor of materia medica, Dr. John B. Beck, began to show the symptoms of a deep-seated malady which greatly affected his strength. For the next three years he suffered from frequently recurring attacks; until he became a confirmed invalid, able to perform his duties only by the exercise of great perseverance and determination. He was at last completely disabled by the progress of his disease, which terminated fatally in 1851.

Dr. Beck was a graduate of the College in the class of 1817. His inaugural dissertation on *Infanticide*, published the same year, was everywhere regarded as an accurate, judicious and exhaustive treatise, showing the early bent of his talent toward medical jurisprudence. In 1822 he was associated with Drs. Dyckman and Francis in establishing the *New York Medical and Physical Journal*, of which he was for several years the chief editor. This journal was very successful. It took the place of its venerable predecessor, the "Medical Repository," and was continued in 1830 under the name of the "New York Medical Journal." Dr. Beck devoted to it much of his time and contributed many articles to its pages. In 1829 he delivered at the College an introductory lecture on the *Higher Departments of Education;* in 1830 one on the *Analysis of the Study of Medicine;* and in 1839 a valedictory to the graduating class on the *Means of Professional Eminence;* all showing an enlightened and hearty appreciation of the dignity of medical art. In 1842 he was president of the Medical Society of the State of New York, his address on the occasion being a *History of American Medicine before the Revolution;* a valuable and interesting record, illustrating the early growth of our medical legislation, modes of practice, hospitals, medical education and medical literature. In 1849 he published his most important work, *Essays on Infant Therapeutics*, which was received with great favor and appeared after his death in a second edition, in 1855. His Lectures on *Materia Medica and Therapeutics* were printed in 1851, under the editorship of Dr. Gilman, and went through two subsequent editions.

In his own department Dr. Beck was "impregnable;" his knowledge being both extensive and accurate. As a teacher his reputation was early established, and continued to increase throughout his life.

He was most earnest and efficient in support of the College during the time of its adversity, and always prompt in its defence against cavil or aggression. The salient points of his character are given in a feeling memorial, written by his friend and colleague, Dr. Gilman. He combined energy of action with steadiness of purpose; and his clearness of perception was equally remarkable in scientific and practical affairs. He was uncompromising in his aversion for fraud or pretension of every kind, and outspoken in the manifestation of his disapproval. His loss, both as professor and associate, was deeply felt by all connected with the College.

On the death of Dr. Beck his chair was filled by *Dr. Elisha Bartlett*, a graduate in medicine of Brown University, Rhode Island. He was a man of singularly engaging manners and great facility of expression, both in writing and speaking. The early part of his professional life was passed as a practitioner in Lowell, Massachusetts, where he became universally popular. In 1836 he was elected mayor of the city, and afterward, for several sessions, a member of the Massachusetts legislature; in both of which positions he was conspicuous for his readiness and fluency in public addresses. This accomplishment, added to his professional knowledge, gave him superior qualifications as a teacher; and at the time of his connection with the College of Physicians and Surgeons he had lectured with success in five or six medical schools in different parts of the country. He delivered the course on Materia Medica and Medical Jurisprudence for the session beginning in 1851, and was duly appointed professor of these branches in 1852.

His writings gained a wide reputation for their attractive style and the variety and abundance of their information. Some were addresses on occasional topics; others, dissertations more or less philosophical in character. He was the author of an introductory lecture on the *Objects and Nature of Medical Science*, Lexington, Kentucky, 1841; an essay on the *Philosophy of Medical Science*, Philadelphia, 1844; and an *Inquiry into the Degree of Certainty in Medicine*, Philadelphia, 1848. The work usually considered as his best was that on the *History, Diagnosis and Treatment of Typhoid and of Typhus Fever*, Philadelphia, 1842. It was reissued in 1847, and again in 1852, in an enlarged form, as the *History, Diagnosis and Treatment of the Fevers of the United States*. On commencing his course at the College as professor in 1852, he delivered an introductory lecture on the *Times, Character and Writings of Hippocrates*. This lecture, which was published by the class, is one of his most characteristic productions; containing the evidence of much learning and research, relieved by frequent touches of playful description and poetic fancy.

But Dr. Bartlett was able to continue the labors of his professorship for only a short time. He was already suffering from a painful affection, which increased in severity during the following year; and in

1853 he was compelled to relinquish active occupation. In the hope that he might be restored to health and usefulness, his colleagues for two years divided his duties among them; but his condition became steadily worse, and in May, 1855, he transmitted his resignation to the authorities of the College. His death took place in July of the same year.

This was the occasion of a partial redistribution of subjects among the remaining professors. In 1847 an additional chair had been created in the College, entitled the chair of "Physiology and Pathology." Neither of these subjects under the arrangement then existing, could be taught with sufficient completeness to meet the requirements of the time. Pathology was wholly unprovided for, except as each professor might give it his incidental attention. Physiology was hardly more than an appendage to the chair of anatomy; to be treated in a cursory way, by stating the function of an organ after the description of its anatomical structure. But such a combination of the two most essential elementary branches was now far from satisfactory; and the professor of anatomy and physiology could no longer perform his double duty without slighting in some degree either one subject or the other. It was deemed almost indispensable, for the continued improvement of the College, that physiology should be separated from anatomy; and furthermore, that a new department should be established, embracing the subjects of physiolgy and pathology.

Accordingly, on the recommendation of the faculty,

the Board of Trustees addressed a memorial to the Regents of the University, in the following terms:

"The subject of physiology is now confided to the professor of anatomy. It has been found, however, that anatomy is required to be taught so minutely, to meet the wants of the students, that it is impossible for the same professor to do justice to the other branch. On the other hand, the recent application of improved microscopes to healthy and diseased structures, together with the great advances in the department of animal chemistry, and the light it has shed on the constitution of our bodies in health and disease, and upon healthy and disordered functions, leave an hiatus in these departments so great that, in the unanimous opinion of the Board, a new professorship is required."

The chair of Physiology and Pathology was therefore created, and in 1847 *Dr. Alonzo Clark* was appointed Lecturer on these branches. Dr. Clark was a graduate of the College in the class of 1835. While still a pupil he had served as assistant in the department of chemistry; and since his graduation he had been almost constantly occupied with pathology and the microscopic examination of healthy and morbid tissues. His known proficiency in these respects made him peculiarly fitted for the new professorship; and he had also shown the necessary talent for imparting instruction. He had lectured at the College for several years in the Spring course; and since 1842 he had been professor of pathology in the Berkshire Medical Institution at Pittsfield, Massachusetts, and in the medical college at Woodstock, Vermont.

After his appointment in the College of Physicians and Surgeons he performed the duties of lecturer for the following session with entire success. Toward the end of the term the president of the College, Dr. Stevens, informed the Trustees that the lectures in the new department "had been regularly given, and had been listened to with great interest;" and that Dr. Clark had "fully realized the high expectations which were entertained of his ability." He was immediately recommended to the Regents of the University, and early in 1848 was appointed professor of physiology and pathology.

This proved a valuable enlargement of the curriculum, and was noticed as follows in the college catalogue for 1852. "The course on Physiology and Pathology has been a very important addition to the regular course of instruction, and is the only course of the kind given in this country. The Lectures on Physiology embrace the minute anatomy of the Tissues, and are amply illustrated by magnified drawings, and by frequent demonstrations under the microscope. The course on Pathology is equally full, and is constantly enriched by the exhibition and demonstration of recent specimens illustrating the various changes produced in tissues and organs by disease."

But after some years spent in this professorship, Dr. Clark's attention was drawn to the more practical departments of medicine. He assisted his colleagues during the illness of Dr. Bartlett, in 1853 and 1854, by delivering a part of the course on theory and practice; and when the chair of materia medica was vacated by Dr. Bartlett's resignation in 1855 a rearrangement of the professorships took place. Dr. Joseph M. Smith became professor of " Materia Medica and Clinical Medicine;" and Dr. Clark was made professor of "Pathology and Practical Medicine." At the same time physiology assumed a more independent position than before, under the name of " Physiology and Microscopic Anatomy;" and this chair was filled by the appointment of *Dr. John C. Dalton*, who had acted as lecturer in the session of 1854-5.

In the same year Dr. Torrey resigned the professorship of chemistry. He was led to this mainly by the pressure of his engagements in the United States Assay Office, of which he had been appointed superintendent in 1853. The lectures on chemistry for 1855-6 were given by Dr. John Le Conte; but in the following year the duties of this chair were performed by *Dr. Samuel St. John*, who was already a teacher of this branch in the medical college at Cleveland, Ohio. He was regularly appointed professor of chemistry in 1857.

Finally the president, DR. ALEXANDER H. STEVENS, resigned his office in November, 1855. He had attended the first course of lectures at the College in 1807; at which time he was eighteen years of age, and a pupil of Dr. Miller, the professor of the practice of physic. He became a member of the Board of Trustees in 1820; professor of surgery at the reorganization of the College in 1826; and at the time of his resignation he had served for twelve years as president. He was for twenty-two years visiting surgeon to the New York Hospital while it was the only institution of its kind in the city, and afterward consulting surgeon to both the New York and Bellevue hospitals. Throughout his active service he was prominent for his popularity and success as a clinical teacher. According to his biographer, Dr. Adams, none of those who heard him could fail to be impressed with his "peculiar aptitude for this department, the kindness of his manner toward the suffering, his avoidance of unnecessary manipulations, his accuracy of diagnosis and felicity of illustration." In the lecture room his language was "familiar, but emphatic and impressive." As an operator he was "cautious, deliberate, and full of resources in unexpected complications." His judgment and skill were eminently practical, and could always be relied on in times of difficulty or danger. In the cholera epidemic of 1832, which caused at its height over one hundred deaths in a day, and produced in the city a wide-spread panic, he was appointed by the Board of Health president of a Special Medical Council for the supervision of all public sanitary matters, holding daily sessions throughout July and August. Of this council Dr. Stevens was the "master spirit;" directing its operations and preparing its documents and reports.

He was one of those who called the first meeting for the formation of the New York Academy of Medicine in 1847. He was elected president of the American Medical Association at its annual meeting in 1848. He was president of the Medical Society of the State of New York in 1849 and 1850; and of the New York Academy of Medicine in 1851.

His earliest work was a translation of Boyer's *Treatise on Surgical Diseases*, New York, 1815. From 1818 to 1841 he contributed a variety of articles on surgery, surgical anatomy, and midwifery to the *Medical Repository*, the *Medical and Surgical Register*, the *New York Medical and Physical Journal*, and the *New York Journal of Medicine and Surgery*. In 1837 he published a clinical lecture on the *Primary Treatment of Injuries;* in the same year *Lectures on Lithotomy*, given at the New York Hospital; and in 1847 he delivered the *Valedictory Address* at the College, a discourse full of sound, practical advice to the graduating class.

His address as president of the State Medical Society in 1849 was a *Plea of Humanity in Behalf of Medical Education*, intended to arouse interest on this subject in the profession and the legislature. It gave an extended review of the benefits conferred on society through the labors of medical men; including inoculation and vaccination, the establishment of hospitals, the improved treatment of the insane, the instruction of the blind and dumb, applications of science to the arts and agriculture, and the diminution of the death-rate by remedial and hygienic means. In the following year his address was on *Public Health*, containing suggestions for First, a sanitary survey of the State, to determine the nature and extent of the existing sources of preventable disease ; and Secondly, the formation of a State medical bureau, to consider matters of legislation affecting the public health.

On the retirement of Dr. Stevens from the presidency of the College, the Board of Trustees adopted resolutions testifying their "deep regret" at his resignation. It was also "*Resolved*, that the interest he has uniformly shown in the welfare of the College, his incessant vigilance in watching over its affairs, and the earnest zeal with which he has either originated or forwarded every improvement in its organization, have secured to him the respect and good will of the Board over which he has so long presided; also, *Resolved*, that, cherishing these sentiments toward their late president, the Board is unwilling to part with him without expressing their continued interest in his welfare, and the hope that he may realize, in retiring from public life, the peace and happiness which he has so richly earned in a long life of professional eminence and active exertion for the good of his profession and the public at large."

Long after his withdrawal from office, Dr. Stevens continued to evince a warm interest in the College; often visiting it during the lecture season, and attending its inaugural or commencement exercises. His presence on these occasions was welcome to young and old. His noble features, with their sedate but kindly expression, and the childlike simplicity of his disposition, combined with firmness and dignity of character, marked him as the possessor of superior qualities and won from his associates the general tribute of their esteem. He died in 1869, at the age of eighty years.

Soon after Dr. Stevens' resignation DR. THOMAS COCK was appointed president of the College. He was already seventy-three years of age, having graduated from the former medical school of Columbia College in 1805; and for nearly thirty years he had been vice president of the College of Physicians and Surgeons. He was visiting physician to the New York Hospital from 1819 to 1834, and president of the New York Academy of Medicine in 1852; and he was interested in various public philanthropic institutions. His character was marked by an unvarying equanimity, and by faithfulness and punctuality in the performance of his duties. But the infirmities connected with his advanced years interfered with his continued activity in the affairs of the College; and he resigned the presidency in 1858. His successor was Dr. Edward Delafield, the former professor of obstetrics.

CHAPTER VI.

REMOVAL TO TWENTY-THIRD STREET.

UNION WITH COLUMBIA COLLEGE.

1856–1860.

By this time the College had been transferred from Crosby street to a new location at the corner of Twenty-third street and Fourth Avenue. Reasons like those which had caused its previous removal were now again operative; namely, the upward movement of the resident population and the adjacent growth of commercial and manufacturing establishments unfavorable to the business of the college. The matter was first broached in 1854, in a communication from the faculty to the Board of Trustees, by whom it was referred to a special committee. It appeared evident to the friends of the institution that such a change was already desirable, and likely to become before long " imperatively necessary." But it involved many difficulties of a financial nature. The Crosby street property was so far encumbered that its sale would leave little to be realized for a building fund; and the cost of a new site and structure must be provided for mainly from other sources. These difficulties were at last overcome, for the most part through the zeal and energy of Dr. Parker and the coöperation of his colleagues. Early in 1855 several vacant lots, embracing seventy-five feet on Twenty-third street by about one hundred feet on Fourth Avenue, were selected as the location for the College. The land was purchased by Dr. Parker for $35,000; and on the fifth of May this purchase, together with building plans and estimates, was considered and approved by the Board of Trustees. At the same time the committee on this subject was authorized to dispose of the Crosby street property, and to apply its proceeds toward the cost of the new establishment. The work proceeded accordingly, and by the end of the year was so far completed that it was determined to remove the College during the winter holidays. The removal took place, and the new building was inaugurated, with an address from the president, Dr. Delafield, January 22d, 1856.

The building was of brick and brown-stone, four stories in height, nearly one hundred feet deep on Fourth Avenue and sixty feet wide on Twenty-third street; leaving on its easterly side a court yard fifteen feet in width, for light and air. The first story was occupied by stores opening on Fourth Avenue and rented for business purposes. On the Twenty-third street front a double flight of stone steps led to the main entrance on the second floor. This story was occupied by a lecture room, forty-five feet by fifty; with the laboratory of the chemical professor in the rear, and various offices and private rooms on the westerly side.

The third floor contained the anatomical museum, forty feet by eighteen; the amphitheatre, forty-five feet by fifty, to seat an audience of a little over three hundred; with private rooms, and waiting and examination rooms for clinical patients; and anatomical and physiological preparation rooms. In the fourth story, which was lighted only from the roof,

THE COLLEGE BUILDING IN TWENTY-THIRD STREET.
1856–1887.
From the Annual Catalogue for 1863.

was the general dissecting room, accommodating twenty-five tables. The building was warmed by a hot air furnace in the cellar for the halls and lecture rooms, and by stoves and fire places for the private rooms. Gas lights and Croton water were generally distributed.

The transfer of the College to its new quarters was not effected without strenuous exertion on the part of the faculty. The expense of the building and its furniture was $55,930; making the total cost of land and building over $90,000. To meet this charge but little more than $9,000 remained in the hands of the Trustees after the disposal of the Crosby street property. Dr. Parker advanced $60,723, secured by bond and mortgage; and the remaining sum of $21,000 was supplied by the professors, in the form of a loan without interest. Much care was needed to secure the success of the enterprise, with due adjustment of the various interests involved. It was finally accomplished; although many years elapsed before the college finances were reduced to a more simple and satisfactory condition.

Soon after the removal to Twenty-third street, the faculty of the College was as follows:

THE FACULTY IN 1858.

EDWARD DELAFIELD, M.D., *President.*

JOSEPH M. SMITH, M.D., *Professor of Materia Medica and Clinical Medicine.*

ROBERT WATTS, M.D., *Professor of Anatomy.*

WILLARD PARKER, M.D., *Professor of the Principles and Practice of Surgery.*

CHANDLER R. GILMAN, M.D., *Professor of Obstetrics and the Diseases of Women and Children.*

ALONZO CLARK, M.D., *Professor of Pathology and Practical Medicine.*

JOHN C. DALTON, M.D., *Professor of Physiology and Microscopic Anatomy.*

SAMUEL ST. JOHN, M.D., *Professor of Chemistry.*

For the next seven years the faculty remained the same. After that time it began to suffer changes by death and resignation; but its membership was not completely renewed until twenty-five years had elapsed.

In 1860 the College of Physicians and Surgeons was made independent of the Regents of the University, and became the *Medical Department of Columbia College*. The former of these changes was desirable on account of certain difficulties in the practical management of affairs, which had become more troublesome with the increasing growth of the College. All the immediate business of the institution was transacted in the city of New York. Everything connected with the courses of instruction, the examination of candidates, the commencement exercises, the selection of proper persons to be recommended for professors or lecturers, or for new members of the Board of Trustees, with other matters of incidental importance, required the personal attention of the faculty and trustees, who were directly responsible for the result. But the ultimate authority for all such proceedings was in the Regents of the University of the State of New York. The assent of this body was necessary for the issuing of each diploma, for the appointment of every new professor or trustee, and even for the validity of by-laws adopted by the trustees for their own guidance.

This complicated governmental machinery often

gave rise to serious inconvenience. The Regents
held their meetings in the city of Albany; and
though abundantly willing to further in every way
the interests of the College, their concurrence in
measures requiring promptitude or punctuality could
not always be obtained in due season. Every year,
after the examination of candidates for graduation, the
list of those found competent by the faculty and recommended by the Trustees must be transmitted to
Albany, the diplomas there made ready and authenti-
cated, and then sent to New York, to be signed by
the president and professors before they could be delivered to the graduates. There were frequent delays
and occasional failures in the transmission of these
documents; and to provide against this contingency
the examinations were sometimes commenced several
weeks earlier, thus virtually abridging the term, so far
as the graduating class were concerned, for purposes
of instruction.

In 1859 the matter was brought before the faculty
and referred to a committee; by whom, early in the
following year, a report was presented embodying
the above facts. A committee of conference from the
Trustees was also appointed, and the subject considered by the joint committee. In accordance with their
recommendation a memorial was forwarded to the
Regents soliciting such amendment of the charter as
should give to the Board of Trustees the power of
final action in regard to graduations, appointments,
and by-laws. There was no opposition to this scheme

on the part of the Regents; but in their opinion it would be of somewhat doubtful legality unless sanctioned by the legislature. It was accordingly referred to that body; and an act was passed March 24th, 1860, amending the charter of the College as follows:

"Section I. The right reserved to the Regents of the University, to confer degrees and to appoint the professors or teachers in the several branches of medical science in the College of Physicians and Surgeons in the city of New York, and of filling all such vacancies as may arise among the trustees or members thereof, are hereby granted to and vested in the Trustees of the said College; and the by-laws which shall from time to time be made by the said Trustees shall be valid and effectual without being confirmed or approved by the said Regents."

By this act the College was relieved from the restraints of its subordination to the Board of Regents, and enabled to transact its business with greater facility. But some explanation seems necessary to account for the ready concurrence given to this change, and for the willingness of the Regents to entrust to other hands a part of their vested authority; especially as on a former occasion their assent had been refused to a similar but less important request. As early as 1825 the Trustees of the College had presented a memorial praying that they might be granted the power of making their own by-laws; and the committee of the Regents to whom the matter was referred had reported that, in their opinion, it would be "inexpedient to abolish or release the right reserved to the Regents on the subject of the by-laws of the College."

The reason for the different view taken by the Regents in 1860 is to be found in the changed condition of both the University and the College. The University of the State of New York, created by the legislature in 1784, was originally intended to include under its supervision and patronage all the higher educational establishments in the State. In 1791 it was empowered to institute, as soon as it might be advisable, a College of Physicians and Surgeons, and to appoint professors or teachers therein; and for twenty years previous to 1835 it was the only recognized authority in the State for conferring medical degrees. But after that time medical colleges began to be incorporated directly by the legislature; and in 1860 there were already a number of such institutions independent of the Regents, and empowered to grant diplomas under the authority of their own Trustees.* The Regents were therefore less solicitous for retaining a control which had so largely diminished in importance; and on the other hand, the College no longer derived from its connection with the University the same prestige as before. Both parties, moreover,

* The names of these Colleges, with their dates of incorporation, were as follows: Medical Institution of Geneva College, 1835; Medical Department of the University of the City of New York, 1837; Albany Medical College, 1839; Medical Department of the University of Buffalo, 1846; New York Medical College, 1850; Long Island College Hospital, 1858.

were ready to adopt any measure which promised to be of practical benefit to the institution.

Soon after the above change had been effected negotiations were set on foot looking to a union of the College of Physicians and Surgeons with Columbia College, as the medical department of that institution. Under the charter which it already possessed, Columbia College was competent to establish such a department; and after several conferences between the respective Boards of Trustees the object was accomplished. On the fourth of June the Trustees of Columbia College formally adopted the College of Physicians and Surgeons as its medical department; and on the sixth of the same month the Trustees of the College of Physicians and Surgeons passed concurrent resolutions, providing that the diplomas for the degree of Doctor in Medicine should be " signed by the presidents of the respective colleges and by the faculty of the College of Physicians and Surgeons," and should be "publicly conferred by the president of the College of Physicians and Surgeons, sitting with the president of Columbia College." A communication from the president to the faculty, dated June 18th, 1860, officially announces that the foregoing action has been taken, and that "the alliance between the two Colleges is now complete."

By this means a union was effected between the two oldest and most prominent institutions in the State for general academic and professional education. In their earlier history they had already been connected with each other. When the medical lectures of Columbia College were suspended in 1813, it was only that the same professors might continue their courses under the new organization; and for more than ten years thereafter they constituted the faculty of the College of Physicians and Surgeons. In 1860, four of the trustees of Columbia College were also trustees of the College of Physicians and Surgeons; and since then the two governing Boards have nearly always had a number of their members in common. Their official connection is thus strengthened by a community of interest in the two institutions, and by personal acquaintance with the affairs of each.

CHAPTER VII.

THE COLLEGE IN TWENTY-THIRD STREET.

1856–1887.

During the thirty-one years passed by the College in Twenty-third street it exhibited many signs of progress and development. Most of them consisted in the continuation or enlargement of improvements already inaugurated, which experience had shown to be of practical utility. Others were the outgrowth of recent times, bringing with them greater opportunities and more exacting requirements.

One of the important events of this period was the formation of the *Association of the Alumni*. Early in 1859 an invitation was issued to the graduates of the College in the following terms.

"Sir: The College of Physicians and Surgeons, instituted in 1807, has now been in existence fifty-two years, and numbers more than eighteen hundred graduates. Among them are to be found men of all ages and distinguished in every department of the profession. It has been thought, by some of the alumni, that an Association, comprehending both the graduates of the institution, and those who have been connected with it, either in its Faculty or as medical members of the Board of Trustees, would be eminently desirable by occasionally bringing together those who from feeling or association have a common interest in its welfare.

"Beside its good effects in bringing together those who have pursued their studies at the same time, and in promoting good feeling and harmony among the graduates of the College, such an association, properly organized, could not fail to exercise, in a variety of ways, a beneficial influence.

"At the request of some gentlemen who have organized temporarily, and formed the nucleus of an association, Dr. Alexander H. Stevens has consented to deliver an Address to the Alumni of the College at the approaching commencement. As an Alumnus of the College, you are cordially invited to attend, on Thursday evening next, March 10th, at half-past seven o'clock.

"For the purpose of permanently organizing the association, a meeting of the Alumni will be held on Friday, March 11th, at the residence of Dr. Delafield, No. 1 East Seventeenth street.

EDWARD DELAFIELD, M.D.	JOHN TORREY, M.D.
THEO. L. MASON, M.D.	GURDON BUCK, M.D.
JOHN WATSON, M.D.	EDW^D. L. BEADLE, M.D.
BENJ. W. MCCREADY, M.D.	ABRAM DU BOIS, M.D.
THO^S. M. MARKOE, M.D.	W^M. H. DUDLEY, M.D.
CHAUNCEY L. MITCHELL, M.D.	CHA^S. M. ALLIN, M.D.
JOSEPH H. VEDDER, M.D.	HENRY B. SANDS, M.D."

In response to this call about two hundred graduates assembled, formed themselves into an association, and elected their officers; among them an orator for the ensuing anniversary.

For the next five years the proceedings of the association consisted, as above, of an address delivered at the college commencement and an annual meeting at the house of one of the resident members for the election of officers and incidental business. But in 1864 Dr. Delafield, who was then president of the College, offered a prize of One Hundred dollars for the best medical essay to be submitted by an alumnus during the ensuing year; and it was decided to dispense with the anniversary address. This led soon after to the establishment of the "Alumni Prize." At the annual meeting in 1866 it was resolved "that the Association establish, as soon as possible, an annual prize, of One Hundred dollars, for the best essay on a medical or surgical subject;" and that, "in order to make this prize perpetual, the sum of Fifteen Hundred dollars be raised by subscription among the alumni and invested as a fund, the interest of which, at least to the amount of One Hundred dollars, shall be annually devoted to this object."

The project thus proposed was realized at the same meeting; the required amount of $1,500 being presented by Dr. Delafield. At first the prize was appropriately named the Delafield Prize. But in the following year it was resolved that the fund be increased to $3,000 by soliciting subscriptions from the alumni; and that the interest thereof "be offered as an annual prize for the best essay upon some subject connected with medicine or surgery." In accordance with Dr. Delafield's expressed desire it was thereafter designated the *Alumni Association Prize of the College of Physicians and Surgeons*. This enlargement of the fund was recommended by the committee for the reason that it would "encourage among the Alumni a spirit of emulation and devotion to scientific research, that must eventually redound to the honor of the College and the progress of medical art."

To this appeal the members of the association responded with such alacrity that by 1870 the fund had already reached the proposed sum of Three Thousand dollars; by 1878 it was nearly Five Thousand; and in 1886 it amounted to Six Thousand dollars, by contributions from about eighty different members. In 1878 it was supplemented by a bequest from *Mr. Benjamin Cartwright*, a friend and patient of the vice president of the association, Dr. A. N. Dougherty. This legacy amounted to Ten Thousand dollars, one half of which was to maintain an annual or biennial prize. This enabled the association to establish two biennial prizes, of Five Hundred dollars each, to be awarded in alternate years; one of them, the "Alumni Association Prize," open to the alumni of the College, the other, the "Cartwright Prize," open to the profession at large. It was required that the competing essays for these prizes, in addition to a high grade of literary excellence, should possess the merit and embody the results of original research.

The remaining half of the Cartwright legacy was devoted to the establishment of lectures, to be given, annually or biennially, on some topic of novelty or interest for the general profession. These are the "Cartwright Lectures," which are delivered, under the auspices of the association, every two years, alternating with the biennial award of the Cartwright prize.

This was the successful issue of the first enterprise undertaken by the Alumni Association. The second was still more important. In 1873 the association adopted a resolution declaring it to be "a prevailing sentiment among the graduates of the College that its field of usefulness should be extended, to meet the continued advance in medical science;" and appointing a committee of conference with the faculty for the furtherance of that object. Agreeably to the recommendations of the committee, a plan was adopted to raise a fund for the special endowment of a professorship of Pathological Anatomy, and for the establishment of laboratories for experimental instruction in chemistry, physiology and pathology. It was believed by the committee that there could be "no more certain way of advancing the standard of medical education than by furnishing the most ample facilities for the prosecution of studies and investigations in the purely scientific departments of medicine;" and that improvements of this kind were "most urgently needed by the College."

A circular setting forth the above plan was distributed to the alumni in New York and the vicinity with such success that in January of the following year Fifteen Hundred dollars had been received toward the formation of the fund. By 1875 it had increased to over Three Thousand; and in 1877 it amounted to Ten Thousand dollars, all contributed by graduates of the College.

By that time it appeared advisable to reconsider, in some respects, the destination of the fund. Its primary object had been mainly the endowment of a chair of pathological anatomy. But certain changes in the faculty were now in prospect which would obviate the necessity for such an endowment; and it was accordingly proposed that the acquired fund should be utilized directly for the establishment of laboratories of instruction. This was declared to be the "aim of the association," and the wisest object to which its means and energies could be devoted. After full discussion and consultation, the above measure was adopted, according to a plan suggested by Dr. Francis Delafield, who was then adjunct professor of pathology and practical medicine. The sum of Ten Thousand dollars already collected, heretofore known as the Endowment Fund, was now designated the "Laboratory Fund." Of this amount the association appropriated Fifteen Hundred dollars for the equipment of the laboratory; the income of the remainder being assigned toward its support during the year. The laboratory was installed in one of the lower rooms of the College building; Dr.

Delafield acting as its Director for the first three years, and bearing no inconsiderable part of its current expenses. Additional contributions were received from graduates and from the faculty. The enlargement and maintenance of the laboratory became a favorite object with the alumni; and in 1884 the Councillors of the association, at a special meeting, resolved to take immediate action, by further subscriptions, for "placing the Physiological and Pathological Laboratory Fund of the Association upon a basis suited to the requirements of the departments of physiology and pathology."

In this way the Association established itself on a permanent footing, as an element of strength and prosperity for the College. In 1873 it was incorporated, under the name of the "Association of the Alumni of the College of Physicians and Surgeons in the city of New York." The purpose of the association is declared in its constitution to be the promotion of "the interests of the College of Physicians and Surgeons in the work of medical education," and the cultivation of "social intercourse among the alumni." Its spirit is expressed, in the report of a committee of the Councillors for 1881, as follows: "The Alumni are the natural guardians of our Alma Mater; it is their province and their privilege to maintain her honor, to stimulate her energy, and to aid her in becoming the representative of progress in medical education."

This object has been assiduously carried out.

There is no part of the organization of the College, of which it can more justly be proud, than the association of its alumni. Their regard for its traditions, and their interest in its reputation and welfare, form the surest guaranty of its future stability; and their counsel and coöperation must be often effective in guiding its policy and in furthering its designs. On the other hand, the highest honor which can attach to the College comes from the general professional standing of its graduates. This is the real purpose of its existence and the ultimate criterion of its success. Moreover, the College and the Association react upon each other to their mutual advantage. The more faithfully the College performs its part in the work of education, the larger will be its returns in the esteem and attachment of its alumni; and alumni can render no better service for the improvement of the profession than by aiding to raise the character and increase the usefulness of their Medical School.

A second feature of interest introduced into the College at this time was the establishment of *prizes for undergraduates*. The first were two prizes, one of Fifty dollars and one of Twenty-five dollars, offered by the faculty in 1858, to be awarded at each annual commencement for the best graduating theses submitted during the year. Their object was to prevent the graduating class regarding the thesis too much as a formality, and to induce, by the proffered honor of a prize, due care in its preparation, as the only written evidence of the candidate's proficiency.

In 1860 a prize was offered by Dr. Alexander H. Stevens "for the best series of preparations which shall adequately illustrate the anatomy, physiology, and pathology of the Larynx." This prize, which amounted to One Hundred dollars and was open for competition to all, was awarded in 1864.

Two years afterward Dr. Stevens established a permanent fund of One Thousand dollars "for the encouragement and improvement of medical literature." This was known as the "Stevens Triennial Prize Fund." Its administration was entrusted to a committee consisting of the president of the College, the president of the Alumni Association, and the professor of Physiology. Its income was to be awarded, once in three years, for the best essay on the subjects proposed by the committee. This prize was also open for universal competition; and although undergraduates of the College could rarely hope for success in obtaining it, they were not excluded from trial by the terms of the endowment.

About the same time a prize was founded by *Dr. Jacob Harsen*, which afterward proved the most important benefaction so far received by the College. It consisted at first of a gold medal and One Hundred dollars in money, to be annually awarded for the best written report of the clinical instruction in the New York Hospital during four months of the year, which should be prepared and presented by an undergraduate student of the College. Productive securities, sufficient to provide for this prize, were conveyed to the College by a formal deed of trust, May 7th, 1859.

Dr. Harsen, whose family were among the early Dutch settlers of New York, was a native of the city and an academical graduate of Columbia College. He

JACOB HARSEN, M.D.
From a medallion portrait by Karl Müller,
1861.

studied medicine in the office of Dr. Stevens, and received his medical degree from the College of Physicians and Surgeons in 1829. For a few years he engaged in general practice, and in the cholera epidemic of 1832 he served the city as district physician. Although he soon afterward relinquished practice as an occupation, he continued to devote much time to professional charities and associations. He was especially interested in the Northern Dispensary, contributing liberally to its support, and serving for twenty years as one of its managers. He was also a Director of the New York Eye Infirmary, and a Trustee and Councillor of the Academy of Medicine. In all these positions he was capable and zealous in the discharge of his official duties. His retirement from practice never diminished his interest in medical affairs nor his regard for the College in which he had received his professional education. He died in 1862, at the age of fifty-four years.

The special form given to this prize was prompted by a double motive. The New York Hospital was the oldest, and for many years the only, medical charity of importance in the city; and even after the establishment of other similar institutions, it remained the favorite with those whose habits or recollections connected them with the social ideas of a former period. Dr. Harsen, both from his professional and family associations, belonged to this class. He believed that the advantages for clinical instruction afforded by the New York Hospital could be made more useful by attracting to them the special attention of medical students; and, on the other hand, as an alumnus of the College, he wished to testify his regard for his alma mater, and to " increase the spirit of emulation" among her pupils. He therefore offered the prize for clinical reports gathered in the New York

Hospital by students of the College of Physicians and Surgeons.

Finding his plan approved by the Trustees and Faculty, Dr. Harsen enlarged it in 1860, to include two additional prizes for the second and third best

THE NEW YORK HOSPITAL

in 1860.

From a print in Appleton's New York Illustrated, 1869.

reports, to be awarded in the same manner as the first. The three were thenceforward known as the " Harsen Prizes for Clinical Reports." They consisted of a First Prize, of the value of One Hundred and Fifty dollars; a Second Prize, of the value of Seventy-five dollars; and a Third Prize, of the value of Twenty-five dollars. With each prize there was given a "Harsen Medal" in bronze, and a diploma signed by the prize committee.

These prizes were annually awarded, from the date of their establishment until 1869. In that year the New York Hospital vacated the property which it had so long occupied in Broadway, with the prospect of acquiring larger and better accommodations elsewhere. But this object was delayed by unforeseen difficulties until 1877; and in the mean time, there being no "clinical instruction in the New York Hospital," no reports were made and the Harsen prizes were suspended.

Their place, however, was taken by prizes offered, by members of the faculty and others connected with the College, for proficiency in special branches of study or for reports of lectures in special departments. In 1874 there was added the "Joseph Mather Smith Prize," endowed by the relatives and friends of the late Dr. J. M. Smith, in commemoration of his long service as professor in the College. It provided for an annual award of One Hundred dollars for the best essay on the subject of the year, presented by an alumnus of the College or a member of the graduating class.

Contrary to expectation, it had by this time become manifest that most of these prizes, instead of being an advantage to the recipients and their competitors, were having an injurious effect. They were offered by professors and teachers in the College, with the sanction of the faculty; and were therefore fairly considered by the students as legitimate objects of competition. All who possessed a fair share of capacity, industry and ambition were naturally desirous of obtaining one of them; especially as it would be to their relatives and friends a source of no little gratification, as the honorable evidence of their ability and merit.

With every year the desire for this distinction increased. To obtain it, many students devoted to the struggle for a specific prize much time and labor which should have been given to their regular studies. This bad result was produced even by the prizes offered for graduating theses. It became not uncommon for a student to select, as the topic of his thesis, some subject requiring prolonged research in the medical libraries or periodicals, with tabulated statistics and bibliographical references; the whole embodied in a voluminous essay, containing elaborate diagrams or colored drawings by a professional artist, carefully engrossed and expensively bound; the candidate meanwhile hardly having time to attend a lecture or demonstration in the College.

The matter was made worse by the increasing number of these inducements, offered, one after the other, by various teachers; until in one year there were no less than ten such prizes open for competition to undergraduates. Their influence was perceptible in the inferiority of the general examinations; and on this account it was determined to abolish the system in favor of a different plan. In 1876 the faculty adopted and promulgated the following preamble and resolutions.

"Whereas, the experience of several years has made it evident, in the opinion of the faculty, that the offering of prizes to undergraduates for theses, clinical or didactic reports, or proficiency in special examinations, has an injurious effect by diverting their attention from the more necessary and legitimate prosecution of their studies: therefore, Resolved;

I. That the Faculty Prizes heretofore offered for graduating theses are hereby abolished.

II. That all professors and teachers connected with the College are requested to withdraw any prizes they may have heretofore offered for the competition of undergraduates. This request is made under a high appreciation of the liberal and commendable motives which have prompted the offering of these prizes; and solely from the conviction that they are unexpectedly operating to the injury of the medical class.

III. In place of the prizes thus discontinued, the faculty hereby establish three prizes, for the members of the graduating class, for *general proficiency in examination;* namely, a First Prize of $100, a Second Prize of $50, and a Third Prize of $25. The manner of awarding these prizes is to be as follows: The ten members of each graduating class who, in their examinations for the degree of Doctor in Medicine and in their graduating theses, have shown the highest proficiency in all the branches combined, shall each receive a diploma of "Examination Honors," and shall be entitled to appear at a public competitive examination in writing for the " Examination Prizes," which shall be awarded to the three successful competitors, in the order of their merit.

This examination shall be conducted by members of the faculty, and the prizes awarded by a committee of three judges, consisting of the president of the College, the president of the Association of Alumni, and a resident alumnus selected by them."

By this means the industry of the student was retained within its proper channels; and the stimulus of competition, instead of concentrating his efforts on a single topic, was made to increase his general proficiency in the requirements of a medical education.

It happened, however, that a special difficulty was presented in the case of the " Harsen Prizes for Clinical Reports." For several years these prizes had been in abeyance, owing to the suspension of the New York Hospital, to which institution they were restricted by the deed of trust. But the new hospital building in Fifteenth street was now approaching completion; and when it should be again in operation the prizes must be offered and awarded as before. During their suspension the prize fund had greatly increased in amount. The property originally conveyed for this object was in the form of thirty-five shares of stock in the United States Trust Company of New York. By accumulation of the unexpended income, and the privileged purchase of new stock for the credit of the fund, it had increased during this time to fifty-three shares, beside a few thousand dollars in other securities; and owing to the enhanced value of the shares themselves the whole fund now represented a sum of about Thirty Thousand dollars, yielding an annual income of nearly Fifteen Hundred dollars. This was far beyond the original intention of Dr. Harsen, and, if applied to the payment of prizes for clinical reports, would be wholly out of proportion to their real merit, and very undesirable as a temptation to undergraduates.

In this dilemma it was thought best to petition the Supreme Court for authority to divert the accumulated surplus of the fund to such other purposes as would be of advantage to the College and in harmony with the spirit of the original donation. The petition was placed by the Court in the hands of the Hon. Richard O'Gorman as referee; whose report thereon was accepted and confirmed by the Court February 25th, 1879. Under this decision, so much of the fund as will carry out the primary object of the trust is retained for that purpose, and still provides for three "Harsen Prizes for Clinical Reports." The remainder is applied to the prizes for general proficiency in the graduating examinations. They are henceforward known as the "Harsen prizes for proficiency at examination," and are increased both in number and in value. The first prize amounts annually to Five Hundred dollars, the second to Three Hundred dollars; and the third to Two Hundred dollars; while the remaining seven of the ten competitors each receive a prize amounting usually to Twenty-five dollars. This liberal endowment is accordingly employed in the way most beneficial to the graduates of the College.

The time spent in Twenty-third street was also signalized by a remarkable growth of the *College Clinics*. In 1856 there were three clinics per week. They were soon afterward increased to four by an additional surgical clinic established by Dr. William Detmold, whose extensive reputation among the German residents of the city attracted a numerous class of patients for examination and treatment. The rising importance of specialities in medical practice also began to produce a demand for various forms of special instruction; and so led to a corresponding expansion of the clinical department in the College. Beside those already mentioned, there were successively added a venereal clinic, a clinic for the eye and ear, one for diseases of the skin, one for children, one for affections of the nervous system, and one for diseases of the throat; until in 1876 there were ten separate clinics per week in the college building.

This enlargement of the clinical service gave rise to greatly increased requirements for accommodation. When the Twenty-third street building was first occupied it was intended that two rooms on the third floor should be fitted up as hospital wards, for male and female patients who might have occasion to remain some days before or after surgical operations. But it soon became evident that all the space available for such purposes must be devoted to the reception and examination of out-patients; and the plan of hospital wards was for this reason abandoned. Moreover, the clinical professors needed room for the storage of their apparatus and illustrations; and the resources of the college edifice were taxed to their utmost, to provide for these increasing demands.

The same difficulty was experienced by all the professors who relied on demonstration in their method of teaching. This feature of college instruction, which had been adopted some years before, became much enlarged while the institution was in Twenty-third street. Not only in chemistry, but especially in anatomy and physiology, the use of instruments, apparatus and material, for purposes of demonstration, grew to such an extent that the accommodation allotted to these departments became excessively overcrowded, notwithstanding every contrivance for economy of space.

Lastly, there were similar requirements for the Physiological and Pathological Laboratory of the Alumni Association. The laboratory was established, in 1878, in an apartment of the college building which had been previously rented as a store. With certain alterations and improvements it answered the purpose very well for a time. But the number of pupils soon increased to such an extent that the room was no longer sufficient. In 1885 the Director announced, in his annual report to the alumni, that the resources of the laboratory in the way of space were exhausted; more students being in attendance than could fairly be accommodated. To meet this necessity additional classes were established, for instruction after the close of the regular session, in May and June. But the relief afforded by this means was still incomplete; since it appeared, from the report of the following year, that the number of students in attendance was " limited only by the seating capacity of the laboratory."

Thus the want of space became on all sides the most serious difficulty in the college operations. Plans were discussed for constructing an additional story on the existing edifice, for building over the vacant space in the rear, and for connecting the college with the adjacent dwelling house on Twenty-third street. But these schemes all proved on examination to be either impracticable, or likely to involve a large expenditure with little corresponding benefit. It was accordingly determined to postpone the attempt until it might be possible to secure a new building with better facilities in a new location.

A further system of improvements adopted at this time consisted of changes in the mode of *instruction and examination*. These changes were partly in the way of simplification, where the methods in use had proved cumbrous or ineffective. Formerly the candidates for graduation were examined orally by the professors, and if found competent were admitted to another oral examination before the Faculty and Trustees in joint session. This second examination, known as the "Trustee examination," was requisite for the final acceptance of the candidate; since degrees could be conferred only with the assent and by the authority of the Board. But it exacted a double expenditure of time from the professors; and it was a source of additional anxiety to students, from the presence of an unaccustomed audience and the special solemnity of the occasion. After this ordeal was passed there still remained the public "Thesis examination," conducted in the main lecture hall of the College, in which the candidate was questioned on the topics of his thesis and expected to answer any objections which might be made to his statements or conclusions. It was an exercise which consumed much valuable time, and had come to be regarded, for the most part, as an empty ceremonial.

These two customs were abolished, at the instance of the faculty, in 1863 and 1864. It was ordered by the Trustees that the second examination, before the joint boards, be dispensed with, and that "in place thereof a committee consisting of five medical members of the Board of Trustees be appointed yearly, to be present at the examination of candidates by the Faculty, of which examination due notice shall be given by the Faculty to the Trustees." In this way the safeguard of the examinations is still maintained, without the inconveniences of the previous method. It is at the same time equally efficacious with the former examination before the full Board; since it is confided to a responsible committee who are competent to the performance of the duty.

The so-called " Thesis examination " was also dispensed with. But the thesis of each candidate was made part of his general examination, and was held to affect the result according to its good or bad quality. This gave to the written exercise its due value in the examinations, and relieved the College from the burden of a useless formality.

Other changes, of more importance than the foregoing, established a greater precision in the mode of examination, and in the requirements for graduation. In former times the question of accepting or rejecting a candidate was decided among the professors by a majority vote. This method was quite sufficient so long as the examinations were conducted before the assembled faculty; because each professor heard the candidates examined in turn by his colleagues, and could form an idea of their fitness in other departments as well as his own. But subsequently to 1856, owing to the increased number of students, it required too much time to hold the examinations in common; and it became customary for each professor to examine the candidates by himself, appearing afterward at a faculty meeting, to report the result and vote accordingly. Under this plan it appeared that a candidate might sometimes be accepted although lamentably deficient in three of the departments, if barely good enough in the other four.

To provide against this contingency the faculty in 1861 adopted a regulation that in passing upon candidates three negative votes out of seven should be sufficient for rejection; with the limitation that two of these votes should come from the practical departments. In 1876 this limitation was abolished, making it requisite in every case that the candidate should be approved by five out of the seven professors; and in 1879 it was finally resolved that candidates for graduation "must pass satisfactorily in each of the seven departments," that is, in anatomy, physiology, chemistry, materia medica, obstetrics, surgery and practical medicine. Two years before, it had also been made requisite that the candidate should present evidence of having pursued the study of practical anatomy.

It was furthermore determined to provide a more definite plan for estimating and recording a candidate's proficiency in each department, and his general standing on the final result. This was done, in 1876, 1877 and 1878, by adopting a uniform standard of marks, on a scale of eight as the maximum; an aggregate of forty for all examinations, with none of them below four, being essential to graduation. By this means the work of voting upon candidates was both facilitated and improved; and it also gave a ready numerical test by which the ten members of the graduating class entitled to "examination honors" might be impartially selected.

In 1879 two additional changes were made which revolutionized still further the process of examination. By the first, all examinations for the degree of Doctor in Medicine, heretofore conducted orally, were required to be in writing; and by the second, the time for holding these examinations was postponed until after the close of the college lectures.

In the system of written examinations, the candidates are assembled, at an appointed hour, in an apartment assigned for the purpose, and supplied with a series of questions arranged beforehand by the professor. Each candidate is provided with a numbered blank-book, in which he writes his answers to the questions, and which he gives up before leaving the room. The professor then reads the answers, records his judgment thereon in each case by the number of the book, and reports the same to the faculty; the name of the candidate being withheld until the reports of all the professors have been rendered. The influences of personality on the mind of the examiner are thus excluded, and his opinion of the candidate's merit is based solely on the written replies in an examination paper. This is usually considered as of prime importance in competitive examinations; and those for graduation in the College partake of that character, since all the candidates are, or may be, competitors for examination honors.

The postponement of examinations till after the end of the college course was of great practical advantage. Experience had shown that attendance upon lectures was of little avail after the beginning of the examinations. The prospective graduates, undergoing or anticipating their examinations from day to day, were not in a condition to profit by the ordinary methods of college instruction; and they had a well-founded suspicion that lectures delivered in any department, after their examination, could not give them information which they were required to possess beforehand. The excitement of the occasion was also more or less contagious, affecting in some measure the whole of the attending class; and on all sides the improvement was manifest when the regular business of the lecture session was brought to its close without interruption, and the examinations conducted afterward in a more deliberate and effective way.

The foregoing changes for the better operation of the College were accomplished by means of successive trials, each of which had its share in the final result. They covered, in all, a period of eighteen years; and in nearly every instance they were considered and discussed for months before their adoption. They were intended, for the most part, to establish a higher standard of requirements for graduation, and to induce, by the knowledge of these requirements, a more thorough preparation on the part of the student. But this alone would not be sufficient. A College which should rest satisfied with raising its demands upon the candidate by increased severity of the examinations would be obviously overlooking an important part of its duty. It should also enlarge the extent and improve the quality of its instruction; since the legitimate purpose of such an institution is not merely to exclude incompetent persons from the profession, but rather to provide those who enter it with the best opportunities for medical education.

The faculty were fully alive to these considerations. They felt that the annual college course was both too short and too crowded for the proper instruction of the class. This course was originally of four months' duration; namely, from the first of November to the first of March. In 1847 it had been lengthened to four months and a half, from the middle of October to the first of March; and in 1868 it was extended to five months by beginning the lectures on the first of October. But the matter to be treated in all the departments continued to accumulate, until an increased allotment for lectures and demonstrations was again necessary. Moreover, there was a growing belief that much of the instruction was compressed into too short a time, and that it ought to be given in some way less irksome and monotonous for the student. A series of lectures on different subjects, closely followed hour after hour throughout the day, can hardly fail to overtax the memory of the listener; and however assiduous he may be, he cannot, under such a method, acquire and retain all the needed information. This was an additional reason for increasing the length of the college term.

In 1880 the faculty, in accordance with the report of a committee, adopted the following resolutions:

"I. That it will be of great benefit to the students to pursue their studies for seven months instead of, as now, for five months.

II. That most of the chairs need more lectures, to make their courses complete.

III. That it will be for the advantage of the students to attend fewer lectures a day, and to have more time for clinical instruction and laboratory work.

IV. That in future the regular session shall commence on the first of October in each year, and last for seven months."

The graduating examinations had already been deferred until after the close of the lectures, bringing the date of commencement, under the new plan, to about the middle of May. Thus the whole time spent in the necessary work of an annual college course was now seven and a half months, or nearly double its former length.

While the practical work of the College was enlarged in these directions, some of its more formal requirements were curtailed or abandoned. The custom of delivering, in each department, an introductory lecture at the beginning of the course gradually fell into disuse. It occupied time which was becoming every year more valuable for the main topics of instruction; and it was dropped, by one professor after another, as each deemed it expedient. But the general Introductory Address, at the annual opening of the session, survived longer. It was an entertainment in which all were expected to take part; the address being delivered in the main lecture hall of the College, with the professors, officers and trustees assembled on the platform. The waning interest in this exercise became manifest in various ways, and the honor of being selected as the orator of the occasion was more commonly shunned than desired. But though nearly all were in favor of its discontinuance, no one possessed the authority to set it aside. It had the negative vitality of a long-established custom, and it continued to exist by that reason alone. It was finally abolished, in 1884, by an order of the Trustees in accordance with the recommendation of the faculty.

CHAPTER VIII.

CHANGES IN THE FACULTY.

1865-1887.

The changes in the faculty of the College, after its removal to Twenty-third street, began with the death of Dr. Chandler R. Gilman. For nearly two years his failing health had interfered so much with the duties of his profession, that he was induced to give up practice in the city and retire to his country home in Middletown, Connecticut. Here he found relief from the more urgent symptoms of his disorder, with great increase of general comfort; and his death, which took place in September, 1865, was without immediate premonition or physical suffering.

During the twenty-five years of Dr. Gilman's connection with the College, he maintained with his associates the most cordial relations. He was a man of the strongest individuality, great kindness of heart, a quick sense of humor, and a remarkably companionable disposition. While not especially distinguished for laborious industry, he was prompt to recognize the progress of ideas in his own department of medicine. He was an enthusiastic amateur of natural science; and in the field of general literature his taste and acquirements were more cultivated and more extensive than is usual with professional men. His opinions were liberal but pronounced; and he knew how to emphasize their expression by a narrative or a witticism. His ample forehead, his keen and rest-

CHANDLER R. GILMAN, M.D.,
Professor of Obstetrics, 1841-1865.
From an engraved portrait by Ritchie,
1864.

less eye, and the flexible intonations of his voice added to the zest of his conversation and to the impressiveness of his style in the lecture room. The sentiment of friendship was strongly developed in his disposition; and he would spare no pains to serve the memory of a deceased colleague, or to further the interests of his surviving family.

His published works were as follows: *Hints to the people on the prevention and early treatment of Spasmodic Cholera*, New York, 1832; *Introductory Address* to the students in Medicine of the College of Physicians and Surgeons, New York, 1840; Maunsell's *Dublin Practice of Midwifery*, edited, with Notes and Additions, New York, 1845; *Periodic Maturation and Discharge of Ova*, in the Mammalia and the Human Female. Translated from the German of Th. L. G. Bischoff, by C. R. Gilman, M.D., and Theodore Tellkampf, M.D., New York, 1847; *Sketch of the Life and Character of John B. Beck, M.D.*, New York, 1851; Beck's *Lectures on Materia Medica and Therapeutics.* Prepared for the press and dedicated to the Alumni of the College of Physicians and Surgeons, by C. R. Gilman, M.D., New York, 1851; *The Relations of the Medical to the Legal Profession*, an introductory address delivered at the College of Physicians and Surgeons, New York, 1856; *A medico-legal examination of the case of Charles B. Huntington*, with remarks on moral insanity and on the legal test of sanity, New York, 1857; Beck's *Elements of Medical Jurisprudence*, eleventh edition. Revised and edited by C. R. Gilman, M.D., Philadelphia, 1860.

Dr. Gilman's successor in the chair of obstetrics was *T. Gaillard Thomas, M.D.*, who had been professor adjunct since 1864. He graduated at the

Medical College of South Carolina in 1852; becoming soon afterward a resident of New York and acquiring a reputation for unusual aptitude in both the teaching and practice of obstetrics. He was for five years a lecturer on this subject in the Medical Department of the University of the City of New York, and was equally successful in the instruction of private pupils. As professor in the College of Physicians and Surgeons, his superiority became widely known. After some years his attention was turned more exclusively in the direction of gynecology, and his weekly clinic for diseases of women made a most attractive feature of the college course. In 1872 he was relieved of the obstetrical portion of his duties by the appointment of *Dr. James W. McLane* as professor adjunct; and some years later the chair was formally divided, Dr. Thomas becoming professor of Gynecology and Dr. McLane professor of Obstetrics. In 1882 Dr. Thomas resigned the didactic portion of his chair and was made professor of Clinical Gynecology, retaining the charge of his weekly clinic; and in 1885 the department was again completed by the appointment of *Dr. George M. Tuttle* as professor of Gynecology. Dr. Tuttle, who was a graduate of the College in the class of 1880, had already served as assistant to the professor of obstetrics and as lecturer adjunct on gynecology.

The next chair to lose its incumbent was that of materia medica, vacated in 1866 by the death of professor Joseph M. Smith, at the age of seventy-seven years. Dr. Smith had long been the senior member of the faculty, having the unique distinction of a continuous service of forty years as professor in the College. For the following session the lectures in this department were delivered by Dr. Freeman J. Bumstead, who also lectured in the College on venereal diseases. But in 1867–8 the course on materia medica was given by *Dr. James W. McLane* as lecturer; and in 1868 he was duly appointed professor. Dr. McLane was an alumnus of 1864, having thus become professor in the College after the unusually short period of four years from the date of his graduation. He retained this position until 1872, when he resigned it to take part in the course on obstetrics, for which subject he had a strong predilection. He was succeeded in the department of materia medica by *Dr. Edward Curtis*, who was appointed lecturer in 1872 and professor in 1873.

Dr. Curtis graduated in medicine at the university of Pennsylvania in 1864, and immediately afterward entered the medical service of the army of the United States as assistant surgeon. Here he continued for some years, occupied mainly in the microscopical section of the Army Medical Museum at Washington. After resigning his commission he established his residence in New York, where he gave a series of lectures in the College on histology, and was thereupon selected to fill the vacant department of materia medica. He discharged the duties of this chair with distinguished ability for fourteen years, when the pressure of other occupations induced him to withdraw from active connection with the College. His place was taken by *Dr. George L. Peabody*, an alumnus of 1873, who was appointed lecturer in 1886 and professor in 1887. He was already visiting physician to the New York Hospital, and had acquired a high reputation as pathologist in the same institution.

The professor of anatomy, Dr. Robert Watts, died in 1867, after a faithful service of twenty-eight years. His chair was filled by the appointment of *Dr. Henry B. Sands*, an alumnus of 1854, who had acted for some years as Assistant Demonstrator and Demonstrator of anatomy, and as lecturer adjunct on anatomy in the preceding session. He continued in this professorship until 1879, when he was transferred to the department of practical surgery, and *Dr. Thomas T. Sabine* became professor of anatomy. Dr. Sabine was an alumnus of 1864, and, like his predecessor, had served as Assistant Demonstrator and Demonstrator of anatomy. For eight years he had been professor adjunct of anatomy, performing during that time a considerable portion of the duties of the chair. The extent and accuracy of his information, and his ability and originality as a teacher, were manifested in a variety of ways; and the value of his instruction was enhanced by many novel methods and ingenious appliances. It may almost be said that he opened a new field for demonstration in one of the most demonstrative branches of medical science.

The chair of surgery became vacant in 1870 by

the resignation of Dr. Willard Parker. This distinguished character, equally eminent as professor and practitioner, was for thirty years the most notable member of the college organization. Immediately on his appointment in 1840 his influence became felt among his associates, and from that time forward he was a leader in all measures for the support or advancement of the institution. He originated the college clinics in 1841. In company with Dr. Watts he secured for the College, by purchase from its former owner, an important part of the existing anatomical museum. He was instrumental, in 1850, in throwing open to students the clinical advantages of Bellevue Hospital, where he was attending surgeon. He was one of the founders of the Pathological Society and of the Academy of Medicine, and he served as presiding officer in both institutions. He urged and aided, against many serious obstacles, the removal of the College to Twenty-third street; and for twenty years afterward he carried upon his own shoulders the greater part of its financial responsibilities. With a robust and active frame, great mental fortitude, and a hopeful and enterprising disposition, he had the faculty of communicating to others a portion of his own impulsive energy. He was not given to elaborate research, nor versed in the minutiæ of medical literature. But he possessed the ready judgment and intelligent self-reliance of a skilful practitioner; and in cases of doubt or difficulty he was a tower of strength to his patients and his colleagues. In the lecture room his language was simple, direct and appropriate; and its meaning was often reinforced by a homely or expressive illustration. His pupils always retained, in

WILLARD PARKER, M.D.,
Professor of Surgery, 1840–1870.
From a portrait in the possession of the College, painted by Huntington in 1871.

after years, a vivid remembrance of his commanding form, his genial manners, and his sound and practical instruction.

After relinquishing his didactic lectures Dr. Parker

accepted the professorship of Clinical Surgery, which he held for ten years longer, often taking a considerable part in the work of clinical teaching. In 1872 he was elected a member of the Board of Trustees, and afterward vice-president of the College; and to the end of his life he continued to manifest his regard for the institution and his warm interest in its welfare. He died in 1884, at the age of eighty-four years.

Notwithstanding Dr. Parker's resignation there was no interruption of the surgical professorship, which for ten years had been partly in charge of *Dr. Thomas M. Markoe*, as professor adjunct. He was an alumnus of 1841, and had long occupied a high rank in the profession; and on the occurrence of the nominal vacancy in 1870, he was created professor of Surgery. He delivered the lectures of this department until 1879, when the chair was divided; Dr. Markoe continuing as professor of the Principles of Surgery, and Dr. Sands becoming professor of the Practice of Surgery. This arrangement answered for some years the growing requirements of the surgical chair; but in 1886 a further provision was made by the appointment of additional lecturers. *Dr. William T. Bull*, an alumnus of 1872, and attending surgeon to the New York Hospital, was appointed lecturer adjunct on the Practice of Surgery; and *Dr. Richard J. Hall*, an alumnus of 1878, and attending surgeon to St. Luke's Hospital, became lecturer adjunct on the Principles of Surgery. Both these gentlemen discharged the duties assigned them during the following session; and in 1887 they were appointed professors adjunct in their respective departments.

In 1876 the professorship of chemistry changed hands owing to the death of Dr. St. John, who for ten years had performed the duties of the chair with assiduous fidelity. He was devoted to his subject, which he taught in a clear and simple style, according to established methods. His intellectual and moral integrity led him to avoid all doubtful or unnecessary statements; and he never failed to distinguish between his actual and his theoretical knowledge. An interested student of nature in all its departments, he was especially fond of astronomy; and his private observatory, equipped with a transit instrument and an equatorial telescope, was for many years his favorite means of scientific recreation. His modest and truthful nature, and his unselfish disposition, gave him a strong hold on the regard and confidence of his brother professors.

The chair of chemistry was then filled by *Dr. Charles F. Chandler*, who had been professor adjunct since 1872. He was also professor of the same department in the School of Mines of Columbia College, in which he had been actively interested since its organization. He was educated at the Lawrence Scientific School of Harvard University and at the University of Göttingen; and he had received the honorary degree of M.D. from the University of the City of New York. He was thoroughly familiar with the modern developments of scientific and technical chemistry, and could present them to his audience in an especially interesting and instructive way.

In pathology and practical medicine, Dr. Clark was relieved from a portion of his labors in 1876 by *Dr. Francis Delafield*, an alumnus of 1863, and the son of Dr. Edward Delafield, former president of the College. He shared the duties of the chair, as professor adjunct, until 1882, when he received the full appointment of professor in this department.

The chair of physiology was supplemented in 1876 by *Dr. John G. Curtis* as professor adjunct. Dr. Curtis was an alumnus of 1870, and a brother of Dr. Edward Curtis, the professor of materia medica; and he had served as Assistant Demonstrator and Demonstrator of anatomy from 1871 to 1875. He continued in charge of various portions of the physiological course, showing eminent ability in the arrangement and application of experimental methods, until 1883, when Dr. Dalton resigned the professorship and Dr. Curtis was appointed to the chair of physiology.

Beside this renovation in the membership of the faculty, there were other changes, of equal importance, in the distribution of subjects and in the mode of providing for appointments.

Among the most noticeable of these changes was the increase in the number of instructors officially attached to the institution. Owing to the greater variety and prominence of the college clinics it was deemed advisable, in 1869, to establish a new grade of teachers under the name of *Clinical Professors*;

each of whom should be in charge of his special clinic, with such subordinates and assistants as he might require. At first only three such appointments were made; but others were soon added to the list, and in 1883 there were six clinical professors, beside the "professor of clinical gynecology," Dr. Thomas, and the "professor of clinical medicine," Dr. Draper.

Moreover there was a tendency, in the didactic departments, to subdivision of the chairs; gynecology being separated from obstetrics in 1879, and the practice of surgery from the principles of surgery in the same year. This was due to the general movement toward specialization, as well as to the growth of material and the increased requirements for instruction in every branch of medicine. It was a further continuation of what had been slowly going on for many years. In former times widely different subjects had been often entrusted to the same teacher. In the session of 1807-8 the professor of materia medica and botany also lectured on surgery and midwifery; and in that of 1811-12 Dr. Hosack, who was professor of the "Theory and Practice of Physic and Clinical Medicine," also gave the lectures on midwifery and the diseases of women and children. From 1808 to 1814 Dr. John Augustine Smith was professor of "Anatomy and Surgery;" and at the reorganization of the College in 1826 he was appointed professor of "Anatomy and Physiology." In 1847, when the number of professorships was increased from six to seven, physiology was associated with pathology; and even after being recognized as a distinct branch in 1855, it still carried with it for several years the added title of "microscopic anatomy." The occasional redistribution of subjects, which took place during this time, was partly in accordance with the preferences or qualifications of individual professors; but as a rule it was in the direction of further subdivision of the chairs.

For instruction in the dissecting room, beside the Demonstrator of anatomy there were successively added a First and Second Assistant Demonstrator; and the Director of the Physiological and Pathological Laboratory was also provided with assistants, as their services became requisite. There were furthermore, recognized and approved by the College, teachers for private classes in various practical and demonstrative branches. In the annual catalogues, the list of those connected with the regular college instruction increased from nine or ten to more than twenty; and the "clinical assistants," occupied in the necessary work of the college clinics, two in number in 1858, amounted to thirty in 1883.

Lastly it became a practice, during this period, to appoint *professors adjunct*, without waiting for actual vacancies in the faculty. It was usually done at the instance of the regular professor, who named for the position some one whom he believed to possess the requisite ability. If acceptable to his colleagues, the candidate was recommended to the Trustees for appointment as "Lecturer adjunct;" and, if approved at the end of a year, he was appointed "Professor adjunct."

He was then assigned such portion of the lectures as the senior professor might determine, their number being usually increased from time to time ; and on the resignation of the senior professor he became, of course, the most prominent candidate for the chair. Of the nine professors appointed subsequently to 1858, six had already served as professors adjunct for periods varying from two to ten years.

This custom was not established by any formal action of the faculty, and was not considered as obligatory on the professors ; but it had such obvious advantages that it was readily adopted whenever circumstances made it desirable and possible. The annual course of instruction in a great medical school is too important a matter to be left to the daily chances of life and health. The professor in a given department may at any time be disabled by illness or accident, to the temporary interruption of his course ; and either his death or resignation may occur unexpectedly, from causes which are not foreseen. In any event there should be a candidate ready to fill his place, who has the necessary personal and professional qualities, and whose capacity as a teacher is not altogether untried. When such provision has been made beforehand, it adds greatly to the strength of the institution. Moreover, the senior professor can often relinquish with advantage a part of his duties; and the new incumbent has the benefit of a more gradual preparation for his future work.

The president of the College, DR. EDWARD DELA-

FIELD, continued in office from his election in 1858 until his death in 1875; during which time he accomplished much for the advancement of the institution. He effected, in 1860, its withdrawal from the immediate authority of the Regents of the University, and its union on the present basis with Columbia College. He took an active part in forming the Alumni Association. His signature headed the list on the call for its first meeting, to be held at his own residence; and five years afterward he made from his private means the first move toward establishing the Prize Fund of the association. A graduate of the College within the first ten years of its existence, appointed professor at its reorganization in 1826, elected a member of its Board of Trustees in 1839, and administering its affairs as president for the unusually long period of seventeen years, he may be said to have followed its fortunes and taken an interest in its welfare for more than half a century. He was always zealous for its reputation and earnestly devoted to its prosperity.

DR. ALONZO CLARK, the ninth president of the College, was elected in 1875, while still occupying the chair of pathology and practical medicine. During the whole time of his service as professor the growth of his reputation had been uninterrupted; and among the alumni of the College, the hospital internes and the profession at large, he was long regarded as the first consulting practitioner in the city. He seemed to attain this position without striving for it, by the sole influence of his unobtrusive but substantial merit.

With an integrity and impartiality that were proverbial, he was equally well known for his assiduity and precision in the pursuit of knowledge. He was an active member of the Pathological Society, where he surpassed all others in the number and variety of his contributions; and in the Academy of Medicine few speakers could command more respectful and earnest attention. His largest field of activity was Bellevue Hospital, where he was attending physician for thirty years, and where he found his most abundant material for clinical study and instruction. In his methods he was industrious, critical and conservative; resorting to every available source of information, and examining with the same caution the conclusions of other observers and his own. In him, the scientific and practical elements were closely combined; and so long as he continued in the practice of his profession, he never gave up the use of his microscope, his test-tube and his library.

After a service of nine years Dr. Clark found his physical powers inadequate to the formal duties of his position, and in 1884 he resigned the presidency. It was filled by the election of Dr. Dalton, who had retired from his professorship in the previous year.

Owing to the above changes the constitution of the faculty again became a new one, as follows:

THE FACULTY IN 1887.

John C. Dalton, M.D., *President.*
Thomas M. Markoe, M.D., *Professor of the Principles of Surgery.*
Henry B. Sands, M.D., *Professor of the Practice of Surgery.*
James W. McLane, M.D., *Professor of Obstetrics and the Diseases of Children.*
Thomas T. Sabine, M.D., *Professor of Anatomy.*
Charles F. Chandler, M.D., *Professor of Chemistry.*
Francis Delafield, M.D., *Professor of Pathology and Practical Medicine.*
John G. Curtis, M.D., *Professor of Physiology.*
George M. Tuttle, M.D., *Professor of Gynecology.*
George L. Peabody, M.D., *Professor of Materia Medica and Therapeutics.*
William T. Bull, M.D., *Professor adjunct of the Practice of Surgery.*
Richard J. Hall, M.D., *Professor adjunct of the Principles of Surgery.*

And the following Clinical Professors:

T. Gaillard Thomas, M.D., *Clinical Gynecology.*
William H. Draper, M.D., *Clinical Medicine.*
Cornelius R. Agnew, M.D., *Diseases of the Eye and Ear.*
Abraham Jacobi, M.D., *Diseases of Children.*
Fessenden N. Otis, M.D., *Venereal Diseases.*
Edward C. Seguin, M.D., *Diseases of the Nervous System.*
George M. Lefferts, M.D., *Laryngology.*
George H. Fox, M.D., *Diseases of the Skin.*
Robert F. Weir, M.D., *Clinical Surgery.*

CHAPTER IX.

GIFTS AND BEQUESTS TO THE COLLEGE.

1875-1886.

Beside the gifts and bequests already mentioned, made to the College or to the Alumni as special endowments for prize funds or lectureships, there were others of more extensive application and intended for more general purposes.

The first was a bequest from Dr. John McClelland, a former graduate of the College. He was the son of a farmer in Saratoga county, New York, where, during the earlier part of his life, he followed his father's occupation. But he was bent on acquiring an education; and although he only began to carry this design into effect after attaining his majority, he then applied himself to it with such success that in due time he entered at Union College, Schenectady, where he received the degree of A.B. in 1832. After completing his academic course at this institution, he pursued the study of medicine with the same perseverance; graduating at the College of Physicians and Surgeons in 1838, at the age of thirty-three years. Not long afterward he was placed in charge of the city Lunatic Asylum on Blackwell's Island, where he remained four years; and in 1845-6 he was the Resident Physician of Bellevue Hospital, an office including the whole responsibility of the medical superintendence and administration of the hospital as it then was. He subsequently practised his profession in the city, devoting himself to its duties with extreme assiduity for nearly thirty years. He was liberal and considerate toward his patients, and always friendly and generous in his intercourse with younger practitioners. Though never distinguished for a very extensive or lucrative practice, by continuous industry and prudent management of his affairs he gradually acquired a handsome competence; leaving at his death in 1875 an estate of nearly ninety thousand dollars.

Dr. McClelland always retained a warm regard for the institutions where he had received his academic and professional education; and in his will, executed two years before his death, he made bequests of equal value to Union College and to the College of Physicians and Surgeons. The latter bequest was in two equal portions, one of which was given to the College proper, the other to the Association of Alumni; "to be used, in their several discretions, towards advancing professional education by establishing free scholarships, the purchasing of medical apparatus and books for libraries, or aiding in the erection of buildings for their accommodation and convenience respectively." At the final settlement of the estate in 1884 the amount realized from this bequest by the College and the Alumni Association was a little over Fifteen Thousand dollars each.

In 1883 Mr. James T. Swift, of the city of New York, made a gift of Ten Thousand dollars for the equipment and maintenance in the College of a physiological cabinet, in memory of his late brother, *Foster Swift, M.D.*, a graduate of the year 1857. Dr. Swift was one of the most promising men of his time; and with prolonged life and health he would have risen to a high rank in the profession. His amiable disposition, attractive person, and engaging manners, no less than his ready aptitude and quick intelligence, made him a favorite with all, and opened before him a sure prospect of success. He served in the College as assistant to the professor of obstetrics from 1861 to 1865, and became attending physician to St. Luke's Hospital, the Nursery and Child's Hospital, and the Woman's Hospital. He was subsequently clinical professor of diseases of the skin in the Bellevue Hospital Medical College, and lastly professor of obstetrics in the Long Island College Hospital, Brooklyn. It was this department to which he was especially devoted, and in which he showed the highest promise of future eminence. But the symptoms of commencing phthisis compelled him to relinquish its duties; and after resorting to various climates in the hope of regaining his health, he died at the island of Santa Cruz, in 1875, at the age of forty-two years.

His brother, Mr. James T. Swift, wished to establish in his honor a permanent memorial, which should be useful to the College and subservient to medical education and research. With this object he created the fund known as the Foster Swift Memorial Fund; to be used for the purchase and safe-keeping of the " more expensive, delicate, and complicated instruments and appliances, mainly instruments of precision, requisite for the pursuit of physiological science." This collection, designated as the *Swift Physiological Cabinet*, was to be kept in a special apartment, for the use and under the immediate charge of the professor of physiology. As there were then no proper accommodations for such a collection in the college building, the fund was placed at interest until the time should arrive when a new building could be obtained and a suitable apartment appropriated therein for the purpose required.

This object, which had become on all sides a pressing necessity, was realized sooner than was anticipated; for in the following year the College received the most ample donation ever made to a medical school in the United States. In October, 1884, MR. WILLIAM HENRY VANDERBILT gave to the institution the land which it now occupies, comprising nearly half the block between Fifty-ninth and Sixtieth streets, Ninth and Tenth Avenues; and a fund of Three Hundred Thousand dollars for the erection thereon of new buildings. As the cost of the land was Two Hundred Thousand dollars, the value of the whole donation was half a million.

In his letter announcing the gift, Mr. Vanderbilt says:

" I have been for some time examining the question of the facilities for medical education which New York possesses. The doctors have claimed that with proper encouragement this city might become one of the most important centres of medical instruction in the world.

"The health, comfort and lives of the whole community are so dependent upon skilled physicians that no profession requires more care in the preparation of its practitioners. Medicine needs a permanent home where the largest opportunities can be afforded for both theory and practice. In making up my mind to give substantial aid to the effort to create in New York city one of the first medical schools in the world, I have been somewhat embarrassed as to the manner in which the object could be most quickly and effectively reached. It seems wiser and more practical to enlarge an existing institution which has already great facilities, experience and reputation than to form a new one. I have therefore selected the College of Physicians and Surgeons because it is the oldest medical school in the State, and of equal rank with any in the United States.

"I have decided to give to the College Five hundred thousand dollars, of which I have expended Two hundred thousand dollars in the purchase of twenty-nine lots situated at Tenth Avenue and 59th and 60th streets, the deed of which please find herewith, and in selecting their location I have consulted with your treasurer, Dr. J. W. McLane. The other three hundred please find enclosed my check for. The latter sum is to form a building fund for the erection thereon, from time to time, of suitable buildings for the College."

Mr. Vanderbilt's personal history was one of grad-

WILLIAM HENRY VANDERBILT.
From a bust in the possession of the College,
modelled from life by Mr. J. Q. A. Ward,
1885.

ual but continuous development of a high order of business capacity. Nearly twenty years of his early life were spent in the cultivation of his farm on Staten Island, which he enlarged and improved until it yielded a handsome revenue. Becoming interested in the affairs of a local railroad, he proved equally successful in its management; and in 1865 he was adopted by his father, Cornelius Vanderbilt, as aid and adviser in the railroad system under his control. In this wider sphere of occupation his ability was soon manifest to all; and he finally became the successor to his father's business responsibilities, as well as the principal heir to his wealth. Under his steady application and practical judgment both of these were greatly increased; and for many years he exercised a controlling influence in the development and extension of railroad interests in the United States.

In his expenditures he was liberal and unostentatious. With but few tastes for personal indulgence, he wished that whatever he had should be the best of its kind. In his stable were the fleetest trotting horses in the country, and his picture gallery contained many works by the first artists of the century. His simple and large-minded disposition was evinced in both his public and private benefactions. In 1880 he furnished one hundred thousand dollars for the transportation to this country of the Egyptian obelisk now standing in Central Park; and so quietly was it done that the monument was placed in position before the public were aware by whom the necessary funds had been supplied. Four years later he advanced to General Grant a loan of one hundred and fifty thousand dollars, to relieve what was thought to be a temporary embarrassment. But when it turned out to be irretrievable ruin, due to the unsuspected dishonesty of a business partner, General Grant insisted upon giving up to Mr. Vanderbilt all the real estate which he possessed, while Mr. Vanderbilt as persistently endeavored to cancel the debt. Failing to overcome the scruples of his distinguished debtor, he consented to a generous compromise; accepting, in full discharge of the obligation, General Grant's collection of military relics, medals and trophies, which he at once presented to the government at Washington, to be held as the property of the nation.

At the time of making his donation to the College, Mr. Vanderbilt was sixty-three years of age, and had retired in great measure from active business. His family physician was the professor of obstetrics in the College, Dr. McLane, for whom he entertained a strong feeling of friendship, and upon whose judgment in medical matters he was accustomed fully to rely. His first intention was to have given a sum sufficient to rebuild on the location of the Twenty-third street property; but he soon felt that this would be an injudicious policy, and decided to enlarge the gift so that the institution might be removed to some point more in keeping with the continuous growth of the city. Nearly a year was spent in inquiries and examinations with this object before the land at Fifty-ninth street was secured and the deed executed. All the preliminary negotiations were conducted by Dr. McLane, and the property even remained in his hands for six months thereafter, until it could be legally transferred to the College. By a provision in its charter, the institution could hold property only to the amount of One Hundred and Fifty Thousand dollars; and it was necessary to apply to the legislature for an amendment relieving it from this restriction. The amendment was granted by an act passed April 17th, 1885; whereupon the land and the building fund were formally placed in the possession of the College.

But Mr. Vanderbilt was not destined to witness the completion of his munificent enterprise. Though in the apparent enjoyment of health and vigor, his tissues were undergoing the insidious change which prepares them for cerebral hemorrhage; and on the eighth of December, 1885, little more than a year from the date of his benefaction, he was stricken down by an apoplectic attack so sudden and overwhelming that his death was nearly instantaneous. Less than three hours before, he had finished his last sitting for the portrait bust now standing in the main hall of the College.

Not long after the death of Mr. Vanderbilt his plans for the enlargement of the College were extended in additional directions by the surviving members of his family. They appreciated the far-reaching value of an endowment which, by increasing the facilities for medical education, provided for the more intelligent and successful treatment of disease; and early in the following year they founded two new and important institutions, to be engrafted on the college organization.

In January, 1886, a communication was received from Mr. and Mrs. William D. Sloane, the son-in-law and the daughter of Mr. Vanderbilt, proposing to erect and endow, on the college grounds, a lying-in asylum, to be known as the " Sloane Maternity Hospital of the College of Physicians and Surgeons." The plan for this establishment, which was at once approved and accepted by the Trustees of the College, included the construction of the hospital by Mr. Sloane, its transfer to the Board of Trustees as the property of the College, and its endowment by Mrs. Sloane for the maintenance of thirty free beds for its inmates. A plot of land, seventy-five feet by one hundred, on the southwest corner of the college grounds, at Fifty-ninth street and Tenth Avenue, was appropriated for its use. Its administration was entrusted to a board of five managers, consisting of the president of the College, Mr. Sloane or his representative, a member of the Board of Trustees, and two members of the Faculty. The service of the hospital was to be directed by the professor of obstetrics in the College, with a resident staff appointed from its alumni ; and under their supervision the members of each graduating class were to attend in turn upon the cases of midwifery occurring in the institution. This unique establishment was welcomed by all, as a most valuable addition to a department heretofore extremely limited in the means of practical instruction.

During the past twenty years the growth of the *clinics* had been one of the principal reasons why more space and accommodation were required for the College. It appeared likely that in its new location this growth would be still greater; and it became a question whether the clinical department should not have an establishment of its own, apart from the didactic and laboratory instruction in the college building. The advantages of such a plan were obvious; and in April, 1886, it was placed on a practical footing by the four sons of Mr. Vanderbilt, namely, CORNELIUS, WILLIAM K., FREDERICK W., and GEORGE W. VANDERBILT. Together they contributed a sum of Two Hundred and Fifty Thousand dollars for the erection and maintenance, in connection with the College, of a free Dispensary, as a special tribute to their father's memory, and as a fitting testimonial of their sympathy with his designs.

This institution was to bear the name of the "Vanderbilt Clinic of the College of Physicians and Surgeons." It was assigned a location, seventy-five feet by one hundred, on the northwest corner of the college grounds, at Tenth Avenue and Sixtieth street. The building and its appurtenances were to be the property of the College; and its board of management, like that of the Sloane Maternity, consisted of the president of the College, a representative of the donors, a member of the Board of Trustees, and two members of the Faculty. It provided, in the most liberal and permanent way, for the medical and surgical treatment of out-door patients, for the necessary accommodation of the clinical professors, and for all the general and special clinical instruction in the College.

By these means, within two years' time, the opportunities possessed by the institution for the enlargement of its teaching capacity were increased five fold. Beside the provision for more spacious and convenient buildings, the amount of land was so ample as to leave a considerable surplus, over and above what was required for their immediate location. This was regarded as one of the most valuable features of the new domain. It was felt by all concerned that the progress in medical education, which had been going on so rapidly for a quarter of a century, would undoubtedly continue; and that the next twenty-five years would bring with them additional demands, the exact nature of which could not be anticipated. Experience had shown that a building of limited construction, however ample for the purposes of the time, will become insufficient when new methods are adopted, or new departments added to the old. But with part of the college grounds yet unoccupied, the needs of the future can be left to their own development, to be provided for as they may arise in the time to come.

CHAPTER X.

THE COLLEGE IN FIFTY-NINTH STREET.

1887–1888.

Soon after the purchase of the land at Fifty-ninth street, preparations were set on foot for beginning the work of construction. A Building Committee was appointed from the Board of Trustees, to coöperate with a committee of the Faculty and to report upon suitable architectural plans for a new college building.

The matter was at once taken into consideration, and on the eighth of April, 1885, the committee reported a general plan, including the location, size, shape, character of construction and approximate cost of the new building, with a recommendation that the cellar excavation be proceeded with immediately. The building was to front on Fifty-ninth street, extending thence, across the college grounds, to Sixtieth street. Subsequently the location of the Sloane Maternity Hospital was fixed at the corner of Fifty-ninth street and Tenth Avenue, and that of the Vanderbilt Clinic at Tenth Avenue and Sixtieth street; thus providing ample space for light and air around each building.

The college grounds, like most of the land in that

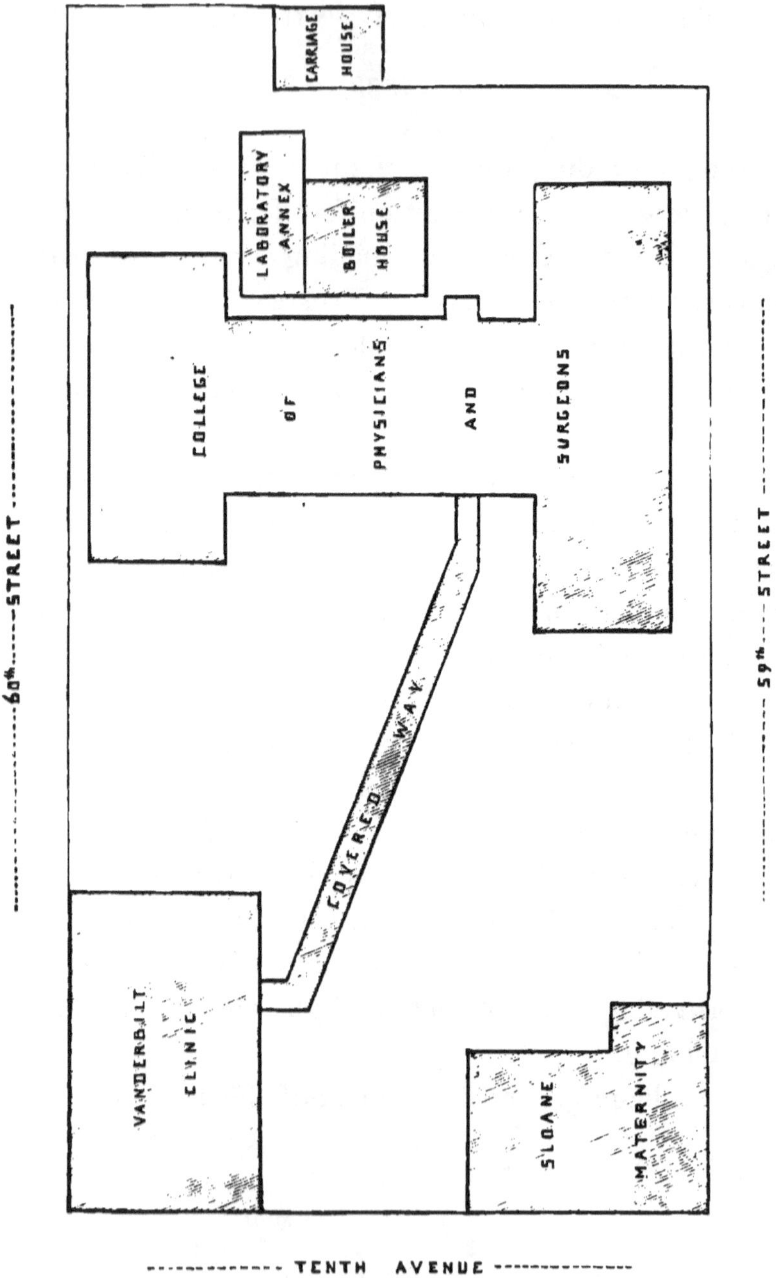

PLAN OF THE COLLEGE GROUNDS AND BUILDINGS

vicinity, consisted mainly of the ledge rock, in many places showing above the surface, and nowhere covered by more than a few feet of earth. This feature of the locality, while it had its advantages, required a considerable expenditure of time and labor in the work of excavation; and it was thought desirable to expedite the construction of the building, by preparing the ground beforehand while the architectural plans were in progress. The recommendations of the committee were adopted by the Board; and the necessary excavation, which included 8,920 cubic yards of rock and 1,257 cubic yards of earth, was begun in May and finished in September, 1885. The cost of this work was a little less than Fifteen Thousand dollars.

Meanwhile the plans for the superstructure of the proposed building were under discussion. Much consultation was required to determine the needs of the various departments and the distribution of space most serviceable for the general purposes of the institution. It was not until January, 1886, that the committee was enabled to report to the Trustees fully detailed plans, with specifications and estimates of cost, suitable for recommendation to the Board. This report and its recommendations were approved and adopted at the same meeting, and the committee was empowered to complete the building contracts in accordance therewith.

Under this authority the foundations of the building were commenced March 19, 1886, the earliest day of the season at which it was thought safe to

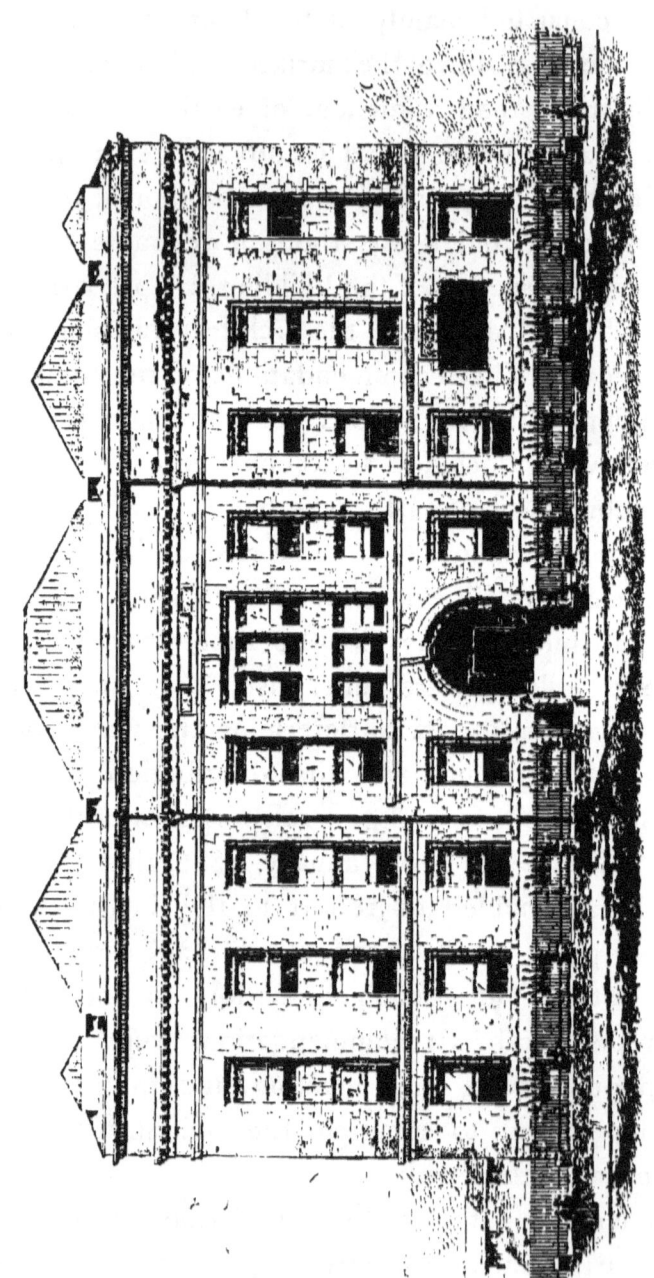

COLLEGE OF PHYSICIANS AND SURGEONS.
59TH STREET FRONT.

begin masonry work. The foundations were completed during the following month; and when the basement walls were about half way up, the corner stone was laid, under the auspices of the building committee, on Saturday, April 24th, 1886.

The exercises on this occasion were opened with prayer from the Rev. Sullivan H. Weston, D.D., Trustee and Chaplain of the College. A leaden box, containing catalogues and other documents relating to the institution, was then deposited in a cavity in the basement wall at the southeast corner of the building, and covered by the granite corner stone, which was lowered into position, guided by the hands of Mr. George W. Vanderbilt. This ceremony was followed by an address from the Hon. Chauncey M. Depew. There was a large attendance, consisting of the Trustees, Faculty, Alumni and other friends of the institution, beside the medical class who were present in a body.

Subsequently the work proceeded without serious interruption; but the substantial character of the building required a liberal allowance of time for its construction. The walls and roof were completed by December. The interior work went on during the succeeding months; and the building was made ready and inaugurated with appropriate exercises on Thursday, September 29th, 1887. The annual college session was opened on the following Monday.

The college edifice consists of three connected structures; namely, a main building, fronting on

170   THE COLLEGE IN

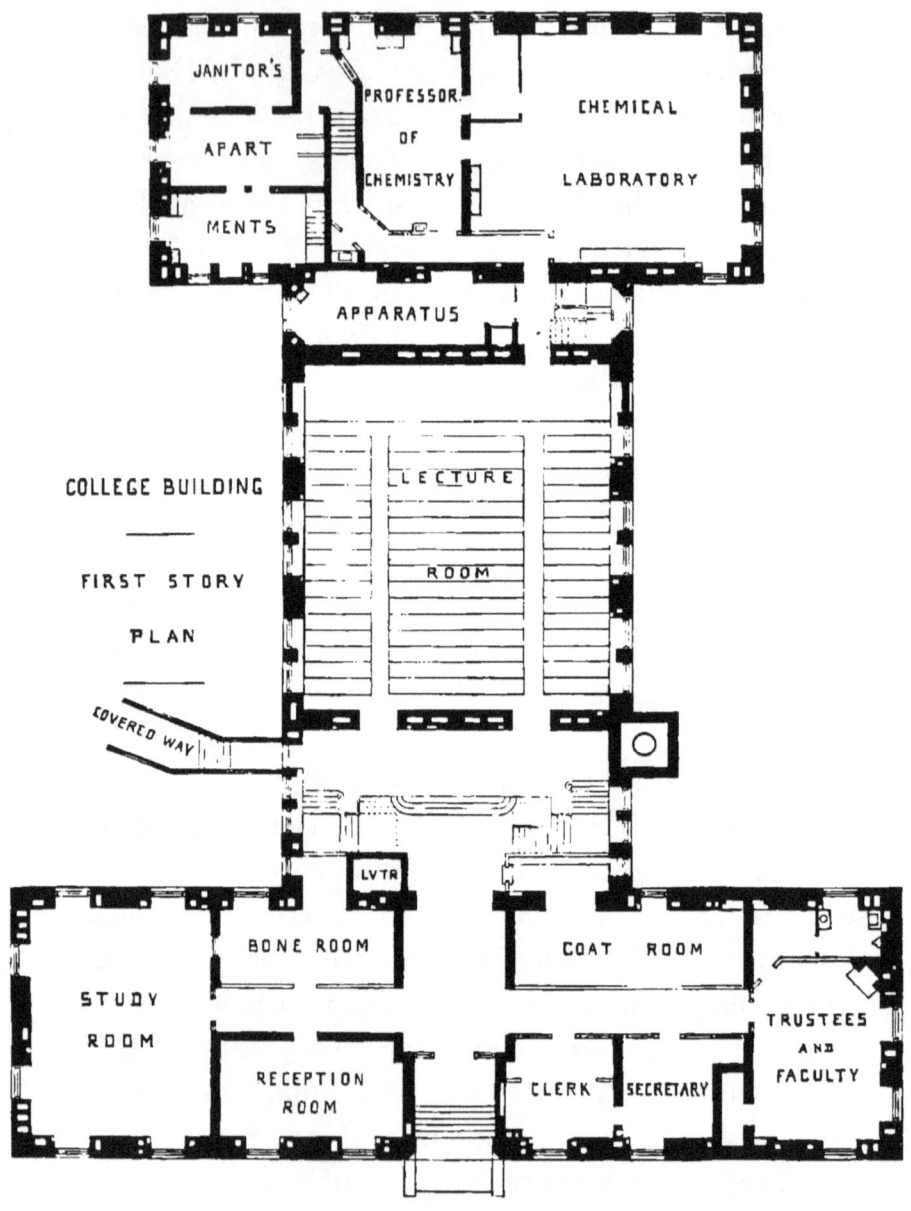

Fifty-ninth street, containing offices, museums, study and recitation rooms, professors' rooms, and the department of practical anatomy; a middle building occupying the central part of the grounds, in which are the main stairway hall, the lecture room, the amphitheatre, and the rear stairway; and a north building or laboratory wing, facing Sixtieth street, containing the janitor's quarters, the chemical laboratories, and the laboratories of the Alumni Association. Outside, and adjacent to the middle building on the east, are the boiler house, and a one-story laboratory annex; and near by a carriage house, with rooms on the second floor for the accommodation of employees. All three buildings are of brick and terra cotta, furnished in various parts with granite sill-courses, lintels, quoins and copings.

The Fifty-ninth street building is one hundred and forty feet long, and forty-three feet deep; of four stories above the basement, and sixty-six feet in height, from the curb to the roof cornice. On the first floor, the main entrance leads into a central hall, sixteen feet wide and fourteen feet high, with transverse corridors extending east and west. Opening on the west corridor are the students' reception room, fifteen feet by twenty-eight, a study room, twenty-eight feet by thirty-six, and a cabinet of osteology, from which students are supplied with parts of the skeleton, or with separate bones, for examination. On the east corridor are the offices of the clerk and secretary, and an apartment for the meetings of the

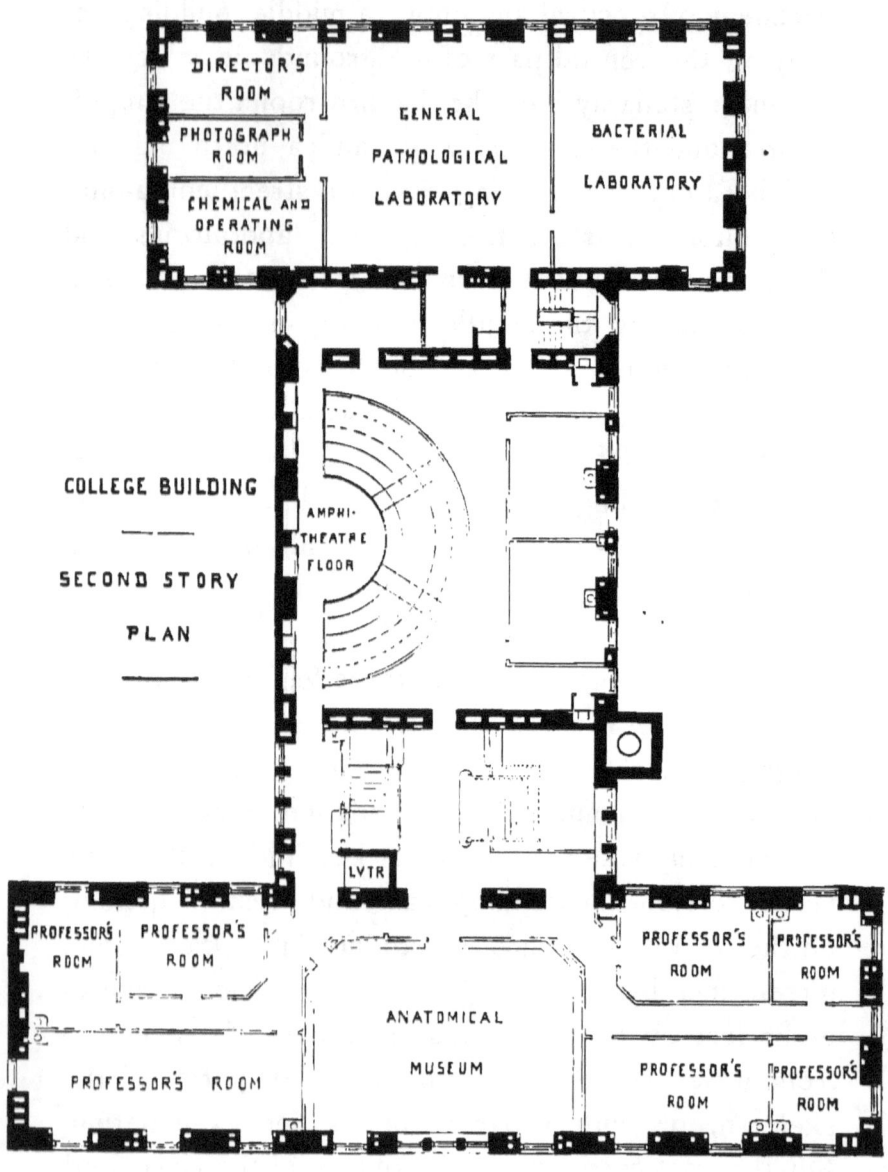

Trustees and Faculty. The door of the coat room opens on the main hall, a little beyond the crossing of the corridors; and immediately adjacent is the stairway leading to the students' toilet room in the basement.

On the second floor is the college museum, a room thirty feet by forty-three, in the centre of the building. The lateral corridors lead to the private rooms of the professors of anatomy, surgery, medicine, obstetrics and gynecology. The third floor is occupied, in the centre by the Swift physiological cabinet, on the west side by the working rooms of the department of physiology, and on the east by rooms for recitation, demonstration and examination. On the fourth floor is the main dissecting room, thirty-six feet by one hundred and five, lighted by three skylights, of which the central one is thirty feet by forty in size, and the two lateral ones each twenty-five feet by thirty. At each end of this floor are additional rooms, with separate skylights, for the demonstrators of anatomy, the prosector of surgery, for private dissecting, and for instruction in operative surgery. At the head of the stairway are rooms for the prosector of anatomy.

The middle building is fifty-five feet in width by ninety-six in length. The first floor contains the main lecture room, forty-eight feet by fifty-five, entered by two doorways from the front hall. It is eighteen feet in height, with a descent of seven feet from the entrance to the front of the lecturer's platform; and is lighted by five windows on each side. Its seating capacity is a little over four hundred. The air-supply

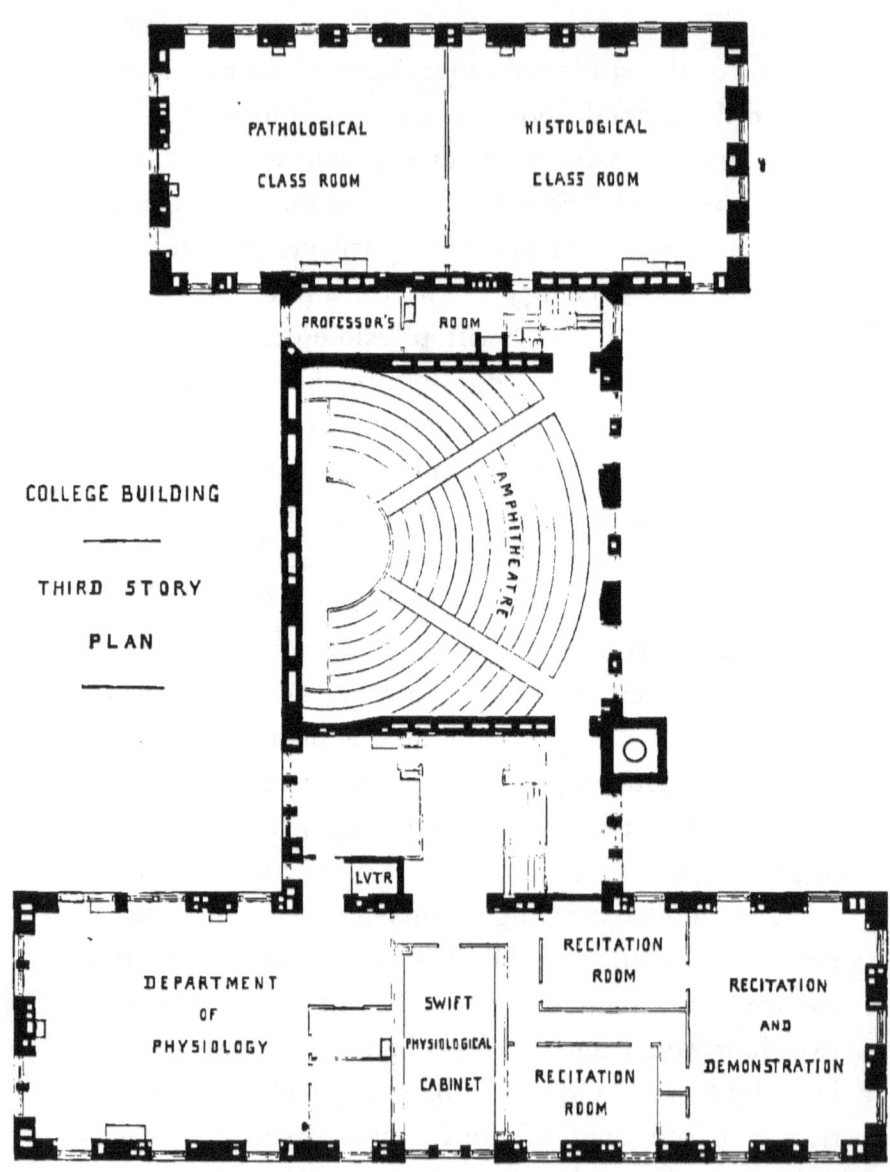

is introduced by narrow horizontal slits, beneath the seats, in the risers of the platforms; and the escape of air is provided for by eight ventilating registers in the north and south walls, four of them being placed near the floor and four near the ceiling. In the rear of the lecture room is an apparatus room for chemistry and physics. In the front hall, at the foot of the main stairway, on the west side, is the entrance to a covered way leading to the Vanderbilt Clinic.

The amphitheatre, situated above the lecture room, is of the same length and width, but has a height of twenty-eight feet from floor to ceiling, with a descent of eighteen feet from the uppermost platform to the lecturer's area. It has a seating capacity of four hundred and fifty. Above the lecturer's area is a skylight, sixteen feet by twenty; and there are six windows in the east wall, behind the upper seats. The air-supply, like that of the lecture room, is by slits in the risers of the platforms; and the air escapes, by openings around the base of the skylight, into the space between the ceiling and the roof, whence it is conducted to the outer ventilating shaft. The professors' entrance to the amphitheatre is by a corridor on the second floor, leading to the area; the students' entrance is by two doors leading, from the front and rear stairways, to the upper platform. Behind the amphitheatre, and communicating with the rear stairway, are two half-story rooms, for the cabinet of materia medica and the private working room of the professor of that department.

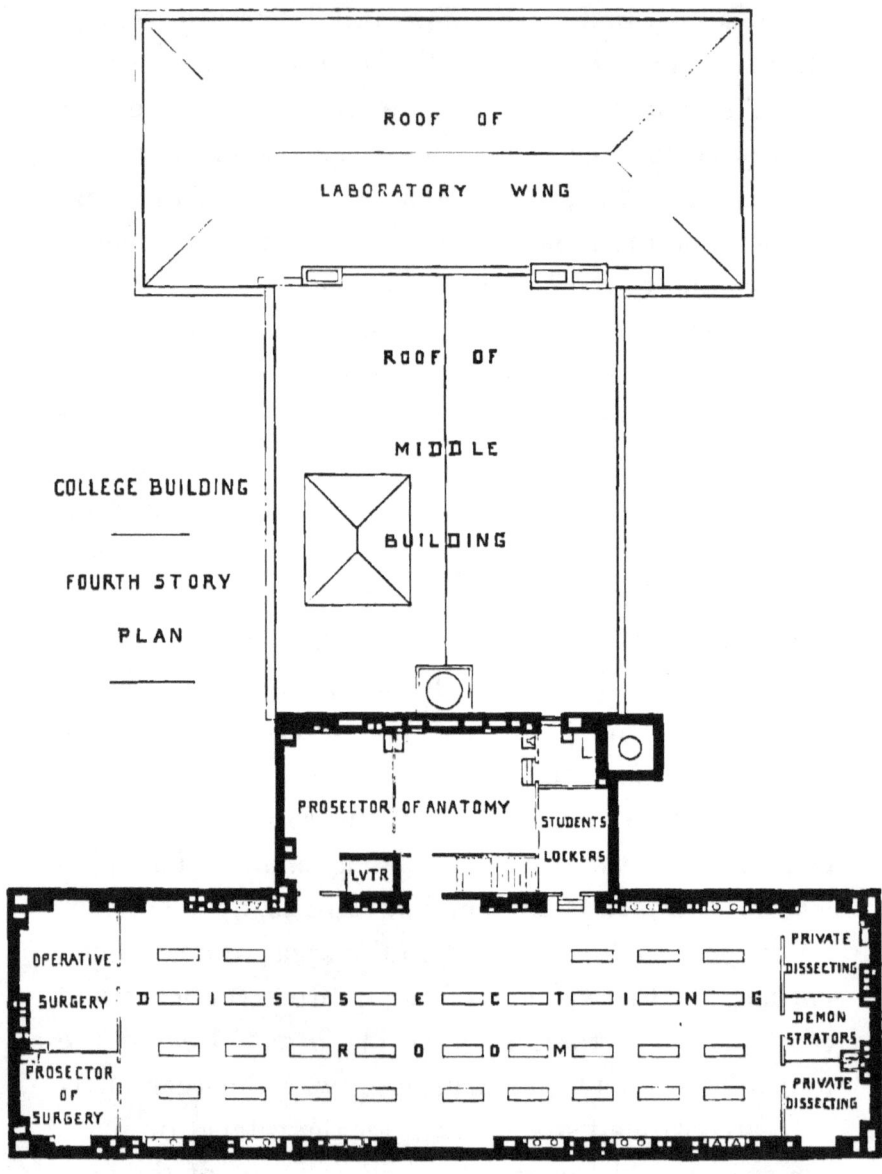

The north or Sixtieth street building is ninety-six feet long by forty-three feet deep, and is three stories in height. On its first floor are the apartments of the janitor, the laboratory of the chemical professor, and a laboratory of instruction in practical chemistry for students. The middle portion of the second floor is occupied by the general pathological laboratory. This is a square room, thirty-six feet by thirty-six, surrounded by shelving for jars of preserved pathological specimens, to be used by the teachers of the laboratory and by advanced workers, for examination and demonstration. Adjoining it on the western end are the director's room, a photograph chamber, and a chemical and operating room. On the eastern end is a room, twenty-seven feet by thirty-six, for experiments and researches in bacteriology. In the northwest corner of this room is an apartment ten feet square, shut off from the rest by glazed partitions, and serving for the experimental culture of various bacterial forms under uniform temperatures. The third floor is divided into two rooms, each thirty-six feet by forty-four, for class instruction in histology and pathology. All the rooms are provided with appropriate tables, hoods, sinks, instruments and apparatus.

Adjacent to the main stairway hall is a hydraulic elevator, running from the cellar to the fourth floor, for the transportation of subjects, specimens, illustrations and apparatus. In the rear hall is a dumb waiter of smaller dimensions, for the accommodation of the laboratory wing.

The warming and ventilation of the building are provided for by five steam boilers, each fifty-four inches in diameter by sixteen feet in length. From these boilers steam is supplied to the heating coils in four air chambers, situated in the basement of the middle building. Four fans and fan-engines drive the air past the heating coils, through as many systems of distributing air ducts, to the rooms above. The first of these systems supplies air to the first, second, and third floors of the front building; the second, to the dissecting room floor; the third, to the lecture room and amphitheatre in the middle building; and the fourth, to the laboratory wing. The air chamber of the third system is provided with valves, by which it may be made to communicate, at will, with either the lecture room or the amphitheatre, or with both. Under a moderate speed of the fan-engine, either of these rooms may be supplied with 750,000 cubic feet of air per hour.

The ventilating flues for the escape of air from all rooms of the front building terminate in two main ducts, which open in the interior of a brick shaft surrounding the iron smoke-pipe of the boiler furnace. One-half the space of this shaft is appropriated to the duct from the dissecting room floor; the other half, to that from the rest of the front building. The ducts from the lecture room and amphitheatre discharge by a common ventilator above the roof of the middle building; and those from the the laboratory wing, by brick shafts above the roof of the rear building.

The front and middle buildings are lighted by three hundred and fifty electric lamps, of sixteen candle power each, run by a dynamo-electric machine and machine engine in the basement. The machine also supplies electricity to an automatic arc light in the lecture room, for magic lantern demonstrations. Illuminating gas and gas burners are distributed in all three buildings.

The cellar excavations for the Sloane Maternity Hospital and the Vanderbilt Clinic were made in the spring and summer of 1886; and the walls of both buildings were well in progress by the early part of December. During the rest of the winter and spring, work was continued at intervals when the weather was sufficiently favorable; and in the following August the roofs were completed. By the end of the year both buildings were ready for use; and they were inaugurated, with an address by Professor T. Gaillard Thomas, M.D., in the lecture room of the College, December 29th, 1887.

THE SLOANE MATERNITY HOSPITAL is at the corner of Fifty-ninth street and Tenth Avenue, with its entrance on Fifty-ninth street. Its dimensions are sixty-five feet on Fifty-ninth street, by seventy-five feet on Tenth Avenue. It is of three stories and an attic. Its exterior architectural effect resembles that of the college building, its brick surfaces being relieved with facings, lines and mouldings of granite and terra cotta.

Internally, its construction is fireproof throughout. The flooring of the halls and the wainscoting of the stairways are of white marble. In the wards the flooring is of vitrified tiles, with a marble base for the side walls. The surfaces of the walls and partitions are in hard finish.

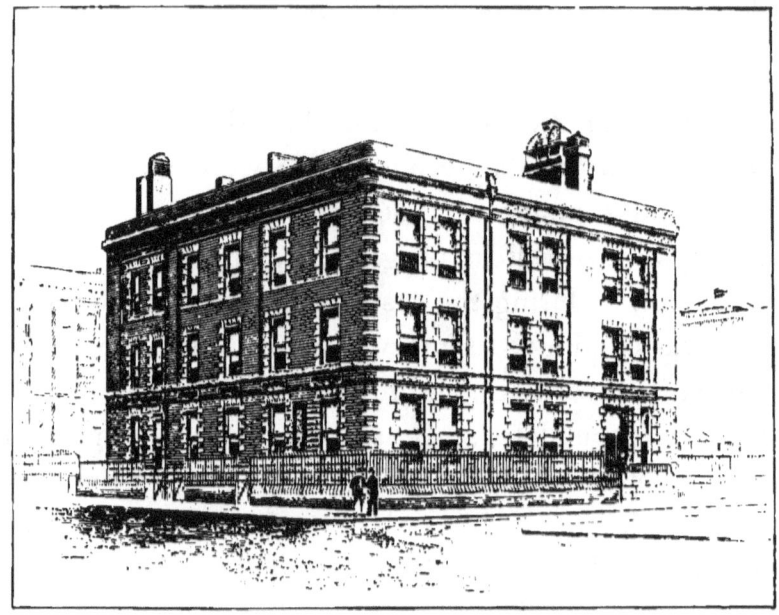

THE SLOANE MATERNITY HOSPITAL,
From Tenth Avenue, near Fifty-ninth street.

The basement contains the laundry, the kitchen, the servants' dining room, the coil chamber and fan for warming and ventilation, and a bath room for newly admitted patients, with lockers for the deposit and safe-keeping of their personal clothing and effects while inmates of the hospital.

On the first floor are the janitor's room, an examination room for patients applying for admission, rooms for the Board of Managers, the House Physician, the Assistant House Physician, and the Matron; also a dining room and a general reception room.

The second floor has three wards, one of six beds and two of four beds each; a delivery room, where the patients are confined and from which they are transferred to the wards after delivery; sleeping rooms for ward nurses; and the drug room of the establishment.

The third floor has three wards, similar to those of the second. It also contains the apartment of the head nurse; sleeping rooms for ward nurses; and two isolating wards, of one bed each, for such patients as may need to be separated from the rest. The attic is occupied by servants' sleeping rooms.

The Matron of the hospital performs the duties of housekeeper; providing for the supplies and other daily requirements of the institution, the hiring and management of servants, and the care and preservation of the furniture. The ward nurses, four in number, are supplied from the training school of the New York Hospital, being selected from those who have already received general instruction for one year or more. Each nurse remains in the Maternity Hospital for three months, learning the duties and performing the service of obstetrical nursing under the instruction of the head nurse, who is permanently attached to the institution and is known as the Principal of the Training School. The House Physician and Assistant

House Physician reside in the hospital, serving for six months in each capacity.

As one of the main objects of the institution is to afford instruction in the practice of obstetrics, a certain number of students belonging to the graduating class of the College are designated each week, to attend the daily visits of the House Physician in the wards, to observe the appearance and treatment of the patients, the dressing and management of the infants, and all matters relating to the puerperal condition. Each of these students in turn also holds himself in readiness to attend, under the direction of the House Physician, the next case of delivery which may occur in the hospital. While engaged in this service they are forbidden to visit any dissecting room or pathological laboratory, or any other hospital or dispensary, or to be present at any kind of anatomical examination elsewhere. The same restrictions apply to the House Physician and the Assistant House Physician during their term of service.

STAFF OF THE SLOANE MATERNITY HOSPITAL
IN 1888.

T. GAILLARD THOMAS, M.D., *Consulting Physician.*
JAMES W. MCLANE, M.D., *Visiting Physician.*
EDWARD L. PARTRIDGE, M.D., *Asst. Visiting Physician.*
JAMES W. MARKOE, M.D., *House Physician.*
HARRY MCM. PAINTER, M.D., *Asst. House Physician.*

The VANDERBILT CLINIC, at Tenth Avenue and Sixtieth street, is similar in general appearance to the two other buildings on the college grounds. Its front on Tenth Avenue has a width of sixty feet, and its depth is one hundred feet. It is three stories in height, with windows in front and on both sides.

THE VANDERBILT CLINIC.
From Tenth Avenue and Sixtieth street.

In the basement are the janitor's kitchen and laundry, a number of store rooms, and the steam apparatus of the building. On the first floor the main hall, paved with white tiles, is used for the accommodation of patients awaiting their distribution to the various clinical departments. On one side of this hall is the drug room, from which prescriptions are dispensed by the apothecary; on the other are clinical rooms for male and female surgical patients, for children, for diseases of the nervous system, and for orthopædics.

The second floor is occupied by clinical rooms for diseases of the eye, the ear, the throat, and for diseases of women. There is also a room with twelve stalls, for instruction in ophthalmoscopy, otoscopy and laryngoscopy; and waiting rooms for patients to be taken into the amphitheatre, the lecturer's entrance to which is on this floor.

In the third story are clinical rooms for practical medicine, for affections of the genito-urinary system, and for diseases of the skin; and a separate stairway for the students' entrance to the amphitheatre. The amphitheatre, situated at the eastern or rear end of the building, is forty-one feet by fifty-seven, horizontal measurement, and thirty feet high from floor to ceiling, with a descent of twenty feet from the uppermost platform to the area floor. Its seating capacity is three hundred and ninety. The air is supplied through slits in the risers of the platforms, and escapes by a large opening in the framework beneath the skylight. The entire building is warmed by steam coil-chambers in the basement; and the ventilation is effected by two fans, one of which serves for the air boxes of the amphitheatre, the other for those supplying the rest of the building.

Both the Clinic and the Maternity Hospital are supplied with steam from the boiler house of the college, through pipes beneath the covered way.

Staff of the Vanderbilt Clinic in 1888.

MEDICINE.

Francis Delafield, M.D., *Professor.*
Frank W. Jackson, M.D., *Chief of Clinic.*
Charles D. Scudder, M.D., *Clinical Assistant.*
George R. Lockwood, M.D., " "
Walter B. James, M.D., " "

SURGERY.

William T. Bull, M.D., *Professor Adjunct.*
Richard J. Hall, M.D., " "
Robert F. Weir, M.D., *Clinical Professor.*
George S. Huntington, M.D., *Chief of Clinic.*
A. J. Magnin, M.D., *Clinical Assistant.*
B. B. Gallaudet, M.D., " "
James R. Hayden, M.D., " "

DISEASES OF WOMEN.

T. Gaillard Thomas, M.D., *Professor of Clinical Gynecology.*
George M. Tuttle, M. D., *Professor of Gynecology.*
Charles Ware, M.D., *Chief of Clinic.*
James B. Hunter, M.D., *Clinical Assistant.*

DISEASES OF CHILDREN.

Abraham Jacobi, M.D., *Clinical Professor.*
Francis Huber, M.D., *Chief of Clinic.*
Abram Brothers, M.D., *Clinical Assistant.*
Albert F. Brugman, M.D. " "
Alexander B. Pope, M.D., " "
Willis W. French, M.D., " "
D. Brown, M.D., " "
H. N. Vineberg, M.D., " "

GENITO-URINARY DISEASES.

Fessenden N. Otis, M.D., *Clinical Professor.*
L. Bolton Bangs, M.D., *Chief of Clinic.*
William K. Otis, M.D., *Clinical Assistant.*
George E. Brewer, M.D., " "

DISEASES OF THE EYE.

Herman Knapp, M.D., *Professor.*
David Webster, M.D., *Chief of Clinic.*
Frank W. Ring, M.D., *Clinical Assistant.*
Walter B. Johnson, M.D., " "
William O. Moore, M.D., " "
John H. Claiborne, M.D., " "
N. J. Hepburn, M.D., " "
Thomas T. Janeway, M.D., " "
Charles H. May, M.D., " "

DISEASES OF THE EAR.

Albert H. Buck, M.D., *Clinical Professor.*
Huntington Richards, M.D., *Chief of Clinic.*
Gorham Bacon, M.D., *Clinical Assistant.*

DISEASES OF THE THROAT.

George M. Lefferts, M.D., *Clinical Professor.*
D. Bryson Delavan, M.D., *Chief of Clinic.*
Urban G. Hitchcock, M.D., *Clinical Assistant.*
Charles H. Knight, M.D., " "
S. O. Vander Poel, M.D., " "
William J. Swift, M.D., " "
William K. Simpson, M.D., " "
Francis J. Quinlan, M.D., " "
Frank E. Miller, M.D. " "

DISEASES OF THE SKIN.

George H. Fox, M.D., *Clinical Professor.*
George T. Jackson, M.D., *Chief of Clinic.*
Charles A. Kinch, M.D., *Clinical Assistant.*
Frank B. Carpenter, M.D., " "

DISEASES OF THE NERVOUS SYSTEM.

M. Allen Starr, M.D., *Clinical Profesor.*
Henry W. Berg, M.D., *Clinical Assistant.*
Barney Sachs, M.D., " "
James A. Booth, M.D., " "
G. A. Dixon, M.D., " "

The ROOSEVELT HOSPITAL is on the south side of Fifty-ninth street, nearly opposite the college building; and its grounds comprise the entire block from Ninth to Tenth Avenue, and from Fifty-eighth to Fifty-ninth street.

The fund for this hospital was established in 1863 by a bequest in the will of James H. Roosevelt, Esq., of property amounting in value to nearly one million dollars. The property was bequeathed to nine trustees, namely; the Hon. James I. Roosevelt, Edwin Clark, Esq., John M. Knox, Esq., and Adrian H. Muller, Esq., individually, and the following five *ex officio;* the president of the New York Hospital; the president of the College of Physicians and Surgeons; the president of the New York Eye Infirmary; the president of the Demilt Dispensary; and the president of the New York Institution for the Blind. It was to be used, under the direction of this Board, "for the establishment, in the city of New York, of a hospital for the reception and relief of sick and diseased persons, and for its permanent endowment." In 1864 the institution was incorporated under the name of the Roosevelt Hospital. The corner stone was laid in October, 1869; and the building was opened for the reception of patients November 2d, 1871.

The hospital consists of a central administrative building fronting on Fifty-ninth street, with an extension in the rear, flanked by three parallel pavilions, two on the eastern and one on the western side.

The central building, four stories in height, contains offices, reception and examination rooms, the apothecary's laboratory and dispensing room, the superintendent's apartments, rooms for the meetings of the Trustees and the Medical Board, an operating theatre, and two surgical wards for women and children. The rear building is occupied by steam boilers and the

ROOSEVELT HOSPITAL.
From the northeast.

warming and ventilating machinery, the laundry, bakery, kitchen and servants' sleeping rooms.

The pavilion next the administration building on the east is of four stories, containing medical wards for male and female patients, nurses' rooms, and apartments for the house staff; that still farther to the east, near Ninth Avenue, is a one-story building for surgical patients; and that on the west, also of one story, is for the treatment of out-door cases.

The hospital contains altogether nearly one hundred and seventy beds; and for the year 1887 there were admitted and discharged over two thousand five hundred patients.

Staff of the Roosevelt Hospital in 1888.

John T. Metcalfe, M.D., *Consulting Physician.*
Thomas Addis Emmet, M.D., " "
Thomas M. Markoe, M.D., *Consulting Surgeon.*
Robert F. Weir, M.D., " "
William H. Draper, M.D., *Attending Physician.*
Francis Delafield, M.D., " "
William H. Thomson, M.D., " "
J. West Roosevelt, M.D., " "
Charles McBurney, M.D., *Attending Surgeon.*
Frank Hartley, M.D., *Assistant Surgeon*
Richard J. Hall, M.D., " "
George S. Huntington, M.D., " "
George M. Tuttle, M.D., *Gynecologist.*
Charles N. Dowd, M.D., *House Physician.*
Alexander H. Travis, M.D., *Senior Assistant.*
Levi O. Wiggins, M.D., *Junior Assistant.*
William H. Park, M.D., *House Surgeon.*
Frederick J. Brockway, M.D., *Senior Assistant.*
Calvin L. Harrison, M.D., *Junior Assistant.*
Robert A. Sands, M.D., *Surgical Dresser.*

When the College, in 1884, came into possession of its grounds at Fifty-ninth street, the neighboring region to the west and north was a nearly uninhabited waste. On the opposite side of Tenth Avenue there was but one building, a country house which had been unused for several years; and the whole of Sixtieth street on the north, from Ninth to Tenth Avenue, was vacant with the exception of a single dwelling house, formerly surrounded by fields and gardens, and still retaining much of its rural aspect. But nearly at the same time with the excavation on the grounds of the College, preparations were commenced for other buildings near by; and when the college edifice was finished and inaugurated in 1887, it was almost completely surrounded on the adjoining streets by five-story brick and stone structures, leaving only the two country houses above mentioned, to represent the former condition of the locality.

The number of students in attendance at the College had greatly increased during the middle and latter part of its stay in Twenty-third street. The average size of the class, for the ten years previous to 1886, was within a fraction of five hundred. In the session of 1886-87, the last held in Twenty-third street, the class numbered a little over six hundred; and in that of 1887-88 it was eight hundred and nine. Notwithstanding the large proportion, in this class, of pupils in the first year of their studies the laboratories of the College were well attended. In the physiological and pathological laboratory there were during this session one hundred and sixty-five pupils, taking fifty lessons of two hours each. Of this number, one hundred and eight followed the course in normal histology, and fifty-seven that in pathological histology. The laboratory of instruction in practical chemistry, though organized for the first time, was attended by sixty-four students.

The year 1888 was signalized by a further advance in the methods of the College, including important changes in the requirements and courses of instruction.

The first of these changes consisted in the adoption of a system of *entrance examinations.* This had long been considered as a desirable measure in the interest of pupils commencing their medical studies. Those who enter upon a course of professional training, without preliminary academic or college education, are at a great disadvantage in regard to their future progress. In many such cases, even with pupils of fair intelligence and ability, nearly a whole year must be spent in " learning how to study." This is especially true for neophytes in medicine. They have not yet acquired the habits of application and retention necessary for so exacting a study ; and in their capacity for mastering its difficulties they are far behind those who possess beforehand the requisite mental equipment. No doubt such deficiencies may be more or less compensated, in certain instances, by unusual exertion or superior talent. But these cases are wholly exceptional ; and for the great mass of medical students it is beyond question that a due preliminary training is in the long run a permanent advantage, and even essential for complete professional success.

To secure this advantage the student must be required to show evidence of systematic study before he is allowed to enter the medical school. As he cannot be expected to know for himself the necessity of such a requirement, it would seem to be the duty of his instructors to impose it upon him for his own benefit. For these reasons the College adopted a schedule of entrance examinations, making it a prerequisite for matriculation that the student should show a fair proficiency in the English and Latin languages, arithmetic, algebra and geometry; such as is usually demanded for graduation in academies or for entrance in advanced literary colleges. These examinations were required of all intending to graduate, except those who had already gone through with similar or equivalent studies in literary colleges or academies, or schools of science, or who had already graduated in medicine elsewhere. This system was promulgated in the college catalogue for 1887, going into operation for the first time in 1888.

An equally important alteration was the adoption of a *three years' college course,* as the regular method of preparing candidates for graduation. The position of medical colleges in this respect had materially changed since the earlier half of the century, owing to the gradual enlargement of their duties and opportunities. Formerly, the student of medicine was required to pursue his studies for the legal term of three years under the direction of a duly authorized practitioner, who was responsible for the character of his instruction, and for the time spent in his pupilage. To receive the degree of M.D., he must also have attended two full courses of lectures in a regularly chartered or incorporated medical college, the last of them in the college which granted him the degree. When the college sessions were only four months long, and consisted wholly or mainly of didactic lectures, the interval between two courses embraced the greater part of the year; and these intervals were occupied with personal instruction from the preceptor. Thus the function of the preceptor was a most important one in the preparation of the student, while the two courses of lectures were short but interesting episodes in his term of study. But as the college session was lengthened from time to time, it came at last to occupy the greater instead of the smaller part of the year. What was of still more consequence, it began to embrace systematic courses of clinical and practical teaching, such as the private preceptor could not hope to give. With the establishment of college laboratories for instruction in histology, pathology and the like, this difference became more marked than ever; until the former relations between office and college instruction were reversed, the duties of preceptor becoming hardly more than an appendage to those of the college.

As this change was plainly a progressive one, and

likely to be more beneficial if made complete, it was proposed that the College should assume, so far as possible, the entire responsibility of preparing its students for graduation. But this could only be done by the adoption of a curriculum extending over the whole time of study; and so arranged as to carry the student, in regular order, from one year to another, beginning with the general and elementary branches and ending with the more specific and practical.

The way had already been prepared, in some measure, for this innovation. For many years past an increasing number of students had actually attended three college sessions; taking, for the first year only the lectures in the elementary departments, following the whole in the second year, and devoting the third to practical subjects and clinical work. This plan, so obvious in its advantages, was pursued, under the advice of the professors, by most of the students residing in the city, and by those from elsewhere who were willing to attend three college sessions instead of two. It was believed that by making such a course obligatory on all, the college would serve, at the same time, its own reputation, the interest of its graduates, and the cause of medical education.

It was evident, moreover, that such a plan should include, as essential parts of the curriculum, the courses of laboratory, clinical, and practical instruction, which had heretofore been left to the individual option of the student. He would thus be directed as to the time and amount of work which he should give to the different departments, both in the lecture room and elsewhere.

Accordingly a schedule for the three years' course was prepared by the Faculty and adopted by the Board of Trustees, to go into effect from the year 1888. Under this plan the first year is devoted to didactic lectures on anatomy, physiology, physics and chemistry, with practical laboratory work in dissection, normal histology, and physiological and medical chemistry. The second year embraces didactic lectures in all the regular courses, with continued practice in dissection, attendance on the general medical and surgical clinics at the Vanderbilt Clinic, and practical clinical instruction in the same departments. For the third year the student follows didactic lectures in materia medica and therapeutics, pathology and practical medicine, the principles and practice of surgery, obstetrics and gynecology; clinics of the special departments at the Vanderbilt Clinic, with practical clinical work on the same; and practical clinical instruction in obstetrics at the Sloane Maternity Hospital.

To increase the efficiency of this scheme, two additional changes were introduced, with the double purpose of enlarging the students' means of instruction, and of giving to the graduating examinations a share in the clinical as well as the didactic departments.

First, the college session was lengthened by another month; the date of Commencement being extended from the middle of May, as heretofore, to the middle of June. This allowed more time for the many requirements embraced in the studies of each year.

Secondly, the conditions for graduation were made to include examinations on the various clinical studies for which provision had now been made. At the same time the "graduating thesis" was abolished. This exercise had been formerly useful, as affording the only written evidence of the candidate's ability to express his ideas in correct and intelligible language. But since the form of examination had been changed from oral questions and answers to that of written papers, and especially since the adoption of entrance examinations, also in the written form, the graduating thesis had ceased to be of practical benefit, and was only an unnecessary burden on both candidates and examiners. It was therefore dropped as a requisite for graduation, and its place in the schedule of examinations was taken by the special clinical subjects followed by the student in his third year.

INDEX.

Academy of Medicine, 84.
Address, Introductory, to the college course, 20, 134; abolished, 135.
Agnew, Cornelius R., M.D., clinical professor, 152.
Almshouse, in Chambers street, clinical instruction in, 21.
Almshouse, at Bellevue, 79 ; converted into a hospital, 80.
Alumni Association, establishment of, 108 ; prize of, 110, 111 ; incorporation of, 114 ; laboratory of, 113, 126.
Anatomy, lecturers on, 12, 69 ; professors of, 19, 30, 55, 63, 72, 101, 141, 152.
Anatomy and physiology, professors of, 55, 72, 147.
Anatomy and surgery, professors of, 19, 147.
Anatomy, physiology and surgery, professors of, 30.
Anatomy, microscopic, appended to the chair of physiology, 93.
Anatomy, practical, legalization of, 81.
Anatomy, surgical and pathological, professor of, 62.

Bangs, L. Bolton, M.D., chief of clinic, 186.
Bard, Samuel, M.D., president of the College, 29, 39.
Bartlett, Elisha, M.D., professor of materia medica and medical jurisprudence, 88 ; his works, 89 ; his illness and resignation, 90.
Beck, John B., M.D., professor of botany and materia medica, 55 ; of materia medica and medical jurisprudence, 62 ; his works, 86, 87 ; his character, 88.
Bellevue hospital, established, 79 ; opened for clinical teaching, 80.
Bequests, to the College, 111, 153.
Botanic garden, Elgin, 36, 37, 38.
Bruce, Archibald, M.D., registrar, and professor of mineralogy, 12 ; of mineralogy and materia medica, 19 ; his biography, 25.
Buck, Albert H., M.D., clinical professor, 187.
Bull, William T., M.D., professor adjunct of the practice of surgery, 144, 145, 185.

Cabinet of materia medica, presented to the college by Professor Beck, 76.
Cabinet, the Swift physiological, provision for, 156 ; established in the college, 173.
Cartwright, Benjamin, his bequest to the college, 111.
Cartwright lectures, 112.
Cartwright prize, 111.
Chandler, Charles F., M.D., professor of chemistry, 145, 152.
Charter, of the college, granted, 9 ; amended, 11 ; supplemented, 29 ; consolidated, 29 ; amended, 104, 161.
Chemistry, practical instruction in, 192.
Chemistry, professors of, 12, 19, 22, 30, 55, 61, 93, 101, 145, 152.
Clark, Alonzo, M.D., lecturer in the Spring course, 74 ; physician to Bellevue hospital, 81 ; advocates the anatomical bill, 82 ; professor of physiology and pathology, 91, 92 ; of pathology and practical medicine, 93 ; elected president of the college, 150 ; his resignation, 151.
Clinic, the Vanderbilt, establishment of, 163 ; location and construction of, 165, 183 ; staff of, 185 ; instruction in, 196.
Clinics, in the college, establishment of, 77 ; growth of, 78, 125, 162, 163.
Clinical gynecology, professor of, 139.
Clinical instruction, at the New York Hospital, 20, 62, 120, 123 ; at the almshouse, 21 ; at Bellevue Hospital, 80 ; in the College, 77, 78, 125 ; in the Sloane Maternity Hospital, 182 ; in the Vanderbilt clinic, 196.
Clinical medicine, professors of, 30, 147, 152.
Clinical surgery, professors of, 144, 152.
Cock, Thomas, M.D., president of the college, 97.
College of Physicians and Surgeons, its origin, 8 ; its title, 10 ; its charter, 9, 11, 29, 104, 161 ; first course of lectures in, 14, 15 ; removal to Magazine street, 21 ; to Barclay street, 32 ; to Crosby street, 63 ; to Twenty-third street, 98 ; to Fifty-ninth street, 165 ; union of, with Columbia College, 102, 106.
College building, in Robinson street, 15 ; in Magazine street, 21 ; in Barclay street, 32, 33, 35 ; in Crosby street, 65, 66 ; in Twenty-third street, 99, 100 ; in Fifty-ninth street, 168.
College, Columbia, formerly near Park Place, 15 ; medical lectures in, 29 ; professors of, join the College of Physicians and Surgeons, 29, 30 ; College of Physicians and Surgeons becomes the medical department of, 102, 106.

College, Rutgers Medical, 53, 58, 59.
Conkling, Frederick A., advocates the anatomical bill, 82.
Curriculum, of the college, extended over three years, 193.
Curtis, Edward, M.D., professor of materia medica, 140; his resignation, 141.
Curtis, John G., M.D., professor of physiology, 146.

Dana, James F., M.D., professor of chemistry, 55, 56, 61.
Dalton, John C., M.D., professor of physiology, 93, 146; president of the college, 151.
Delafield, Edward, M.D., professor of obstetrics, 55; surgeon to the New York Eye Infirmary, 57; lectures on diseases of the eye, 59; resigns his professorship, 69; president of the college, 97; establishes the Delafield prize, 110; his administration as president, 149, 150.
Delafield, Francis, M.D., originator and director of the laboratory of the Alumni Association, 113, 114; professor of pathology and practical medicine, 146, 152, 185; physician to Roosevelt Hospital, 190.
Delafield prize, 110.
Delavan, D. Bryson, M.D., chief of clinic, 187.
Depew, Chauncey M., orator at laying of corner stone of college building, 169.
Dering, Nicoll H., M.D, trustee and registrar, 61.
Detmold, William, M.D., lecturer in the Spring course, 74; establishes a surgical clinic in the college, 125.
De Witt, Benjamin, M.D., professor of the institutes of medicine, 12; of chemistry, 19, 25.
Draper, William H., M.D., professor of clinical medicine, 147, 152; physician to Roosevelt Hospital, 190.

Emmet, Thomas Addis, M.D., consulting physician to Roosevelt Hospital, 190.
Examination, for the degree of M.D., as first conducted in the college, 27, 127; simplified, 128, 129; made more stringent, 129, 130; required to be in writing, 130, 131; deferred till after the close of lectures, 131; to include clinical subjects, 197.
Examination, for entrance into the college, 192.
Examination honors, 122.
Examination, prizes for general proficiency in, 122, 123.

INDEX.   201

Faculty of the college, in 1807, 12; in 1808, 19; in 1814, 30; in 1825, 52; in 1826, 55; in 1843, 72; in 1858, 101; in 1887, 152.
Fox, George H., M.D., clinical professor, 152, 187.
Francis, John W., M.D., registrar and professor of materia medica, 31, 41.
Fund, alumni association prize, 110, 111; alumni association laboratory, 112, 113, 114; Cartwright prize and lecture, 111; Foster Swift memorial, 156; Harsen prize, 116, 123; Joseph M. Smith prize, 120; Stevens triennial prize, 116.

Gifts and bequests to the college, 76, 110, 111, 116, 120, 153, 155, 156.
Gilman, Chandler R., M.D., professor of obstetrics, 69, 72; lectures in the Spring and Fall courses, 74, 75; establishes clinic for diseases of women and children, 78; his exertions for improvement of the college, 79; physician to Bellevue Hospital, 81; his character and appearance, 136, 137; his published works, 138.
Graduation, first ceremonies of, in the college, 27, 28.
Graduation, examinations for, 27, 127, 128, 129, 130, 131, 197.
Gynecology, separated from the chair of obstetrics, 139.
Gynecology, professors of, 139, 185.

Hall, Richard J., M.D., professor adjunct of the principles of surgery, 144, 145, 152, 185; assistant surgeon to Roosevelt Hospital, 190.
Hamersley, William, M.D., professor of clinical medicine, 30; trustee, 61.
Harsen, Jacob, M.D., establishes the Harsen prize, 116; his biography, 117.
Harsen prizes, for clinical reports, 116, 119, 123; for proficiency in examination, 124.
Hosack, David, M.D., professor of materia medica and botany, 12; of the theory and practice of physic, 30; establishes the Elgin botanic garden, 36, 37; his traits and peculiarities, 39, 40.
Hospital, the New York, in 1807, 16; in 1860, 119; removed to Fifteenth street, 123; clinical instruction in, 20, 62, 116, 123.
Hospital, Bellevue, established, 79; opened for clinical instruction, 80.

Hospital, Roosevelt, foundation of, 188 ; trustees of, 188 ; inauguration of, 188 ; construction of, 189 ; staff of, 190.
Hospital, the Sloane Maternity, foundation of, 162 ; location and construction of, 165, 179 ; inauguration of, 179 ; organization of, 162, 181 ; practical instruction in, 182 ; staff of, 182.
Huber, Francis, M.D., chief of clinic, 186.
Huntington, George S., M.D., chief of clinic, 185.

Inauguration, of the college, in Robinson street, 14 ; in Magazine street, 21 ; in Barclay street, 34 ; in Crosby street, 64 ; in Twenty-third street, 99 ; in Fifty-ninth street, 169.
Inauguration, of the Roosevelt hospital, 188 ; of the Sloane maternity hospital and the Vanderbilt clinic, 179.
Institutes of Medicine, chair of, in the college, 19 ; professors of, 12, 19.
Instruction, in the college, changes in methods of, 127, 133, 193.
Instruction, clinical, at the New York Hospital, 20, 62, 120, 123 ; at the almshouse, 21 ; at Bellevue Hospital, 80 ; in the College, 77, 125 ; at the Vanderbilt clinic, 196.
Instruction, demonstrative, 76, 126.
Instruction, experimental, in chemistry, physiology, and pathology, 112, 192, 196.
Instruction, practical, at the Sloane maternity hospital, 182.
Introductory address, at opening of the college session, 20, 134 ; abolished, 135.
Introductory lectures, 20 ; discontinued, 134.

Jackson, Frank W., M.D., chief of clinic, 185.
Jackson, George T., M.D., chief of clinic, 187.
Jacobi, Abraham, M.D., clinical professor, 152, 186.
Jaques, John D., M.D., trustee and treasurer, 61.

Knapp, Herman, M.D., professor of ophthalmology, 186.

Laboratory, physiological and pathological, of the alumni association, 113, 126, 171, 177.
Lecturers, to serve before becoming professors, 73.
Lecturers, on anatomy, 12, 69 ; on chemistry, 12, 25, 93 ; on obstetrics, 69 ; on pathology, 91 ; on physiology, 91 ; on surgery and midwifery, 12 ; on materia medica, 140, 141.
Lecturers adjunct, how appointed, 148.
Lectures, Cartwright, 112.

INDEX.

Lectures, clinical, at the New York Hospital, 20.
Lectures, in the college, first course of, 14; second course of, 20.
Lectures, introductory, 20; discontinued, 134.
Lectures, Spring and Fall courses of, 73, 74, 75.
Lefferts, George M., M.D., clinical professor, 152, 187.
Legalization, of practical anatomy, 81.

McBurney, Charles, M.D., surgeon to Roosevelt Hospital, 190.
McClelland, John, M.D., a benefactor of the college, 153.
McLane, James W., M.D., professor of materia medica, 140; of obstetrics, 139; advises location of college grounds, 157; conducts negotiations for purchase, 160; visiting physician, Sloane Maternity hospital, 182.
Magazine street, second location of the college, 21; becomes a part of Pearl street, 21.
Markoe, Thomas M., M.D., professor of surgery, 144; of the principles of surgery, 144, 152; consulting surgeon to Roosevelt Hospital, 190.
Materia medica, professors of, 12, 19, 24, 31, 41, 55, 72, 86, 88, 93, 101, 139, 140, 141.
Maternity hospital, the Sloane, 162, 165; construction of, 179; inauguration of, 179; organization of, 162, 181; practical instruction in, 182, 196; staff of, 182.
Metcalfe, John T., M.D., consulting physician to Roosevelt Hospital, 190.
Miller, Edward, M.D., professor of the practice of physic and clinical medicine, 12, 19; joins in guaranty of funds to the college, 15; his professional eminence and character, 24.
Mitchill, Samuel L., M.D., vice president, 12, 19; professor of chemistry, 12; of natural history and botany, 19, 26; his accomplishments and reputation, 24.
Mott, Valentine, M.D., professor of surgery, 31, 40; of operative surgery, and surgical and pathological anatomy, 62; professor emeritus, 63.

New York, in 1807, description of, 16.
New York Hospital, in 1807, 16; in 1860, 119; clinical instruction in, 20, 62, 120, 123; removed to Fifteenth street, 123.

Obstetrics, practical instruction in, 181, 182, 196; professors of, 19, 31, 55, 72, 101, 136, 138, 139.
Ophthalmology, professor of, 186.
Otis, Fessenden N., M.D., clinical professor, 152, 186.

Park Place, formerly Robinson street, 15.
Parker, Willard, M.D., professor of surgery, 68, 72 ; lectures in Spring and Fall courses, 74, 75 ; establishes the college clinic, 77 ; his exertions for improvement of the college, 79; surgeon to Bellevue hospital, 81 ; advocates the anatomical bill, 83 ; his resignation as professor, 141, 142 ; his character and influence, 142, 143.
Pathology, importance of, first recognized, 90, 91.
Pathology and practical medicine, chair of, 93 ; professors of, 93, 101, 146, 152.
Peabody, George L., M.D., professor of materia medica and therapeutics, 141, 152.
Pearl, formerly Magazine street, 21.
Physiology, embraced in the Institutes of Medicine, 19 ; joined with anatomy and surgery, 30 ; with anatomy, 55 ; with pathology, 90, 91 ; with microscopic anatomy, 93.
Physiology, professors of, 30, 55, 63, 72, 91, 92, 93, 101, 146, 152.
Post, Wright, M.D., professor of anatomy, 30, 36 ; president of the college, 39 ; his character and reputation, 42, 43.
Presidents, of the college, 12, 21, 29, 39, 42, 55, 63, 69, 71, 93, 97, 149, 150, 151.
Prize, alumni association, 111 ;
    Cartwright, 111 ;
    Delafield, 110 ;
    Harsen, 116;
    Joseph Mather Smith, 120 ;
    Stevens triennial, 116.
Prizes, Harsen, for clinical reports, 116, 119, 123, 124; for proficiency in examination, 124.
Prizes, special, for undergraduates, 115 ; their ill effects, 120, 121 ; abolished, 122.
Professors, in 1807, 12 ; in 1808, 19 ; in 1814, 30 ; complaints against, 43, 44, 45 ; their replies, 46 ; ineligible as trustees, 48 ; disputes of, with the trustees, 49 ; resignation of, 53 ; new appointment of, 55 ; mode of selecting and appointing, 73, 148, 149.
Professors, of anatomy, 63, 101, 141, 152 ;
    of anatomy and physiology, 55, 72, 147 ;
    of anatomy and surgery, 19, 147 ;
    of anatomy, physiology and surgery, 30 ;

Professors, of chemistry, 12, 19, 22, 30, 55, 61, 93, 101, 145, 152;
    of chemistry and botany, 62, 72;
    of clinical gynecology, 139;
    of clinical medicine, 30, 147, 152;
    of clinical surgery, 144, 152;
    of gynecology, 139, 152;
    of institutes of medicine, 12, 19, 21, 25;
    of legal medicine, 31;
    of materia medica, 31, 139, 140, 141;
    of materia medica and botany, 12, 55;
    of materia medica and clinical medicine, 93, 101;
    of materia medica and medical jurisprudence, 62, 72;
    of materia medica and therapeutics, 152;
    of mineralogy, 12; of mineralogy and materia medica, 19, 24;
    of natural philosophy, 30; of natural history and botany, 19, 22, 30;
    of obstetrics, 139; of obstetrics and the diseases of children, 152;
    of obstetrics and the diseases of women and children, 19, 26, 31, 55, 72, 101;
    of operative surgery and surgical and pathological anatomy, 62;
    of ophthalmology, 186;
    of pathology and practical medicine, 93, 101, 146, 152;
    of physiology, 63, 146, 152;
    of physiology and pathology, 91, 92;
    of physiology and microscopic anatomy, 93, 101, 147;
    of practice of physic and clinical medicine, 12, 19, 24;
    of practice of surgery, 144, 152;
    of principles of surgery, 144, 152;
    of principles and practice of surgery, 31, 55, 72, 101, 141, 144;
    of surgery, 19, 30;
    of theory and practice of medicine and clinical medicine, 72, 147.
Professors, clinical, 146, 152.

Professors adjunct, 148 ;
>of anatomy, 141 ; of chemistry, 145 ;
>of obstetrics, 138, 139 ; of pathology and practical medicine, 146 ;
>of physiology, 146 ; of principles and practice of surgery, 144, 145.

Removal of the college, to Magazine street, 21 ; to Barclay street, 32, 33, 34 ; to Crosby street, 63 ; to Twenty-third street, 98 ; to Fifty-ninth street, 165.

Repository, the New York Medical, earliest medical periodical in the United States, 23 ; succeeded by the New York Medical and Physical Journal, 87.

Rhinelander, John R., M.D., professor of anatomy, 63 ; his resignation, 68.

Richards, Huntington, M.D., chief of clinic, 187.

Robinson street, first location of the college, 15 ; afterward Park Place, 15.

Romayne, Nicholas, M.D., first president of the college, and lecturer on anatomy, 12 ; guarantees funds for the college, 15 ; professor of the institutes of medicine, 19 ; his services and characteristics, 21, 22.

Roosevelt hospital, establishment of, 188 ; inauguration of, 188 ; construction of, 189 ; staff of, 190.

Sabine, Thomas T., M.D., professor of anatomy, 141, 152.

Sands, Henry B., M.D., professor of anatomy, 141 ; of the practice of surgery, 144, 152.

Senatus academicus, of the college, 13, 20.

Session, the college, length of, in 1808, 19 ; in 1847, 75, 133 ; in 1868, 133 ; in 1880, 133, 134 ; in 1888, 196.

Sloane, William D., Mr. and Mrs., establish the Sloane maternity hospital, 162.

Sloane maternity hospital, establishment of, 162 ; location of, 165 ; construction of, 179 ; inauguration of, 179 ; organization of, 162, 181 ; practical instruction in, 182, 196 ; staff of, 182.

Smith, Alban G., M.D., professor of surgery, 68.

Smith, John Augustine, M.D., lecturer adjunct on anatomy, 12, 13 ; professor of anatomy and surgery, 19 ; joint professor of anatomy, physiology and surgery, 30 ; his resignation and removal to Virginia, 56 ; his reappointment as

professor of anatomy and physiology, 55, 56 ; president of the college, 63 ; his character and attainments, 69, 70.
Smith, Joseph M., M.D., professor of theory and practice of physic, 55, 57 ; of materia medica and clinical medicine, 93 ; his long service in the college, 139, 140.
St. John, Samuel, M.D., professor of chemistry, 93, 101 ; his character, 145.
Starr, M. Allen, M.D., clinical professor, 187.
Stevens, Alexander H., M.D., professor of surgery, 55 ; surgeon to the New York Hospital, 56 ; his resignation as professor, 68 ; president of the college, 71, 72 ; his resignation, 93 ; his published works, 95 ; his personal and professional qualities, 94, 96, 97.
Students, in the college, number of, first, second and third sessions, 18; eighth, ninth and tenth sessions, 34 ; in 1820 and 1822, 43 ; after 1826, 59 ; from 1876 to 1886, 191 ; in 1886–87 and 1887–88, 191.
Surgery, professors of, 19, 30, 31, 55, 62, 72, 101, 141, 144, 152.
Swift, James T., establishes the Swift Physiological Cabinet, 155, 156.

Thesis, graduating, 27 ; examination on, 27, 128, 129 ; prizes for, 115, 121, 122 ; abolished, 197.
Thomas, T. Gaillard, M.D., professor of obstetrics, 138 ; of gynecology, 139 ; of clinical gynecology, 139, 147, 152, 185 ; orator at inauguration of the Sloane Maternity hospital and the Vanderbilt Clinic, 179 ; consulting physician, Sloane Maternity hospital, 182.
Torrey, John, M.D., professor of chemistry and botany, 61, 62, 72 ; lectures in the Fall course, 75 ; his resignation, 93.
Trustees, of the college, under its first charter, 10, 11 ; under charter of 1812, 29, 31; professors ineligible as, 48 ; disagreements of, with the professors, 49 ; majority of, to be non-medical, 51 ; members of the board of, after 1826, 61.
Trustee examination, 127, 128 ; abolished, 128.
Tuttle, George M., M.D., professor of gynecology, 139, 152, 185 ; gynecologist to Roosevelt hospital, 190.

University, of the State of New York, grants the college charters, 9, 29 ; authority of, over the college, 10, 27, 45, 46, 102, 103 ; transferred to the Board of Trustees, 104.

Vanderbilt, William H., gives land and building fund, to the college, 156; his letter of announcement, 157; his personal history and qualities, 158, 159, 161.
Vanderbilt, George W., lays the corner-stone of the college building, 169.
Vanderbilt, Cornelius, William K., Frederick W., and Geo. W., establish the Vanderbilt clinic, 163.
Vanderbilt clinic, establishment of, 163; inauguration of, 179; construction of, 183; staff of, 185; instruction in, 196.
Ware, Charles, M.D., chief of clinic, 185.
Watts, John, M.D., president of the college, 55, 56, 63.
Watts, Robert, M.D., professor of anatomy, 68, 72, 101, 141; lectures in the Spring and Fall courses, 74, 75; his exertions on behalf of the college, 79.
Webster, David, M.D., chief of clinic, 186.
Weir, Robert F., M.D., professor of clinical surgery, 152, 185; consulting surgeon to Roosevelt Hospital, 190.

www.ingramcontent.com/pod-product-compliance
Lightning Source LLC
Chambersburg PA
CBHW020833230426
43666CB00007B/1202